81

LURCHERS AND LONGDOGS

Two of the author's lurchers, Hettie and Rummage. They are lurcher–lurcher bred, (by Lucky out of Tarn), and collie ancestry is obvious. Some of the so-called 'Smithfields' may well have looked like Hettie.

LURCHERS

AND

LONGDOGS

E. G. WALSH

THE BOYDELL PRESS

First published 1977
Revised edition 1978
Reissued 1984
Reprinted 1985, 1988

Published by The Boydell Press
an imprint of Boydell & Brewer Ltd
PO Box 9, Woodbridge, Suffolk IP12 3DF

ISBN 0 85115 402 6

Printed in Great Britain by
St Edmundsbury Press Ltd, Bury St Edmunds, Suffolk

Contents

To

D. J. R.

and the other neighbouring farmers and
landowners who, from time to time, very
kindly allow me to exercise their hares with
my dogs; and to Stewart, who sews up the dogs
after the inevitable accidents;
my grateful thanks.

List of Illustrations

Introduction

I have owned and worked longdogs of various breeding in England, India and the Middle East; but when I came to look at the lurcher from the point of view of writing about him I realised that, so far as the past was concerned, most of the lights had been kept very firmly under bushels. There is little mention of him in literature and none, apparently, before 1668. I think that this stems from the fact that most of those who kept and used lurchers could not write, and that many of those who knew of them, and could write, were busy trying to exterminate them. There were similar problems in any complete description of the lurcher up to the 1960s; he was an ephemeral being, with no recorded past, a fleeting present and, in many cases, no future. In only one or two lines were any pedigrees kept, puppies lived or died by survival of the fittest, the lurcher's way of life made him prone to damage that often brought an early end to his career. He had no focal point in the country and, outside the circle of lurcher men themselves, very few knew or cared anything about him.

Then came the first Lambourn lurcher show; I have written the story before, both the start and end of it, but, briefly, in 1971 the first Lambourn show drew an entry of under 100 lurchers and the final show in 1982 drew an entry of over 1,400 lurchers. From that one show in 1971 the number grew to nearly 150 lurchers shows in 1983. What caused this enormous rise in lurcher numbers and why have they become so popular?

One reason is, of course, the 'bandwagon' effect; there are always those who will buy a dog if they think it is becoming popular, whether they know anything about that breed of dog or whether they have any intention or means of using it as it should be used. But in my opinion the two main reasons for the lurcher's present popularity are, firstly, the tremendous increase in field sports of all sorts and particularly of the do-it-yourself kind. It is a great thrill to see the pride of your greyhound – or whippet – kennel come out of slips like a rocket, lead white collar several lengths on the run-up, put in a series of slashing turns and kill handsomely, and to see the judge cantering back, holding out the red handkerchief for your dog. But to many lurcher people there is a greater thrill and satisfaction in seeing a lurcher suddenly stop, nose to wind. She puts down her head, canters back and forwards across the wind and makes a dive at a clump of grass. Up bounces a hare, the lurcher misses it at first snatch and settles down behind it like a police driver waiting for the car he is chasing to make a mistake. The hare crosses the field and after a couple of turns gets through a smeuse in the hedge, while the lurcher flies the rails in a gap and follows the hare down the far side. The hare comes back through the hedge with the lurcher close behind her and makes a split-second check for another dash through. This is what the lurcher has been waiting for: she cuts the corner and catches the hare as it jumps the ditch. She stands for moment getting her breath back then, after a couple of trial lifts to get the balance right, she canters back across the field with the hare held high.

Look at another lurcher: it is eleven o'clock on a dark, windy, night. You switch on the spotlamp and you have only swung it a couple of degrees when two red eyes glow in the grass. You let go of one end of the cord through the dog's collar and off he goes, just outside the narrow beam of light. He turns in and nudges the rabbit up to make it run; it is the first one of the night and he is full of himself, but he will be only too glad to pick up sitters in an hour's time. You follow with the beam, he catches the rabbit half-way to the hedge and turns back to you as you switch off the lamp.

Look at a third lurcher, this time one that is rare, partly because this needs a certain type of dog, and partly because many people are too greedy and want too much all at once. It is midnight and a quarter moon is almost down. You finish hooking the ends of a net over the gateposts. The lurcher, a three quarter collie-quarter greyhound, is waiting for your signal. You tell her to go, she hurdles the gate and streaks off up the right hand side of the field like a competitor in a sheepdog trial. You peer over the gatepost and you can just see her, brown and white, crossing the top of the field, coming in ten yards or so and starting across again. She suddenly changes direction and puts on speed; you can't see a hare in front of her, but crouch down now and keep still. You hear a grunt as the lurcher turns, there is the swish of feet in wet grass, a bang against the bottom rail of the gate and a hare is struggling in the loose pocket of net on the ground.

The second reason for the popularity of lurchers is a simple one. Good lurchers are very intelligent, friendly, beautiful, versatile, functional dogs, fully capable of keeping master in grub; no wonder they are so popular.

When I wrote 'Lurchers and Longdogs' eight years ago, I tried to make some guesses at the lurcher's origins and history; to describe and comment on some of the various crosses from which they were bred; to give some facts and anecdotes about their work; to make some suggestions on breeding, rearing, buying and first aid; and to have a layman's look at the lurcher and the law. Apart from some quotations the views expressed are my own. I expected my views to be challenged, not least my definitions. I quoted from such authors as William Scrope, Charles St John and Sir Samuel Baker. They lived in a more spacious age when there was room and time for life and the English sin of minding other people's business was not so widely practised by those with nothing better to do. These authors knew their dogs and the quarry from years of practical work and their books make nostalgic reading.

Other than re-writing this preface I have made no alterations for this re-issue of 'Lurchers and Longdogs'. Obviously there are some alterations that can be made in the light of experience, of lurcher shows for instance, but most of the book is as true now as when I wrote it. Whilst the word 'lurcher' is loosely used to describe any cross-bred looking greyhound-type of dog I did offer my own definitions of certain terms, and as these have come to be accepted by many lurcher men I repeat them here:

A Lurcher is an intentional cross between a coursing dog (greyhound, deerhound or whippet; my opinion of salukis will be found in Chapter 5) and a working dog (sheep or cattle dog, police dog, hound, gundog or terrier); or the offspring of such crosses which do, sometimes, breed true.

A *Longdog* is the term used to describe any 'longtail' but in this context it is the offspring of a coursing dog and a coursing dog.

A *Norfolk lurcher* was a cross between a deerhound or greyhound and a sheep or cattle-dog. Nowadays it is used to mean any long, or broken-coated lurcher over about 23 or 24 inches.

A *Smithfield* was a drover's dog, used for driving sheep or cattle for many miles across the country to better grazing near London or some other big city or to the cattle markets themselves. These dogs were a *type* and not a *breed*. Droving was not just from Norfolk to London; drovers and their cattle came from Ireland, the Western Isles, the Highlands and Lowlands, from Wales, Yorkshire, from the Midlands and East Anglia. Having delivered the cattle the drovers went home and started again. There was no central registry of dogs and no kennels where pure-bred collies could be bought. The drovers used the dogs that they bred at home or which they bought or swapped on the road. They were hard men and lived hard lives and the dogs had to be harder than the men. Translating the requirement into make and shape one can suggest a tough-minded, rough- or broken-coated dog with an intelligent eye, 23 to 25 inches at the shoulder, 40 to 50 lbs in weight, well-muscled, deep in the chest with well-sprung ribs, running up light in the belly, well-let down hocks, good, sound feet and straight action. In other words a medium-sized working lurcher!

Lurchers are now enjoying a boom that cannot be good for them, but it is a boom which will only stop of its own accord. Lurcher shows can be good fun and good meeting places, so long as they are not taken too seriously. The lurcher is a working dog or he is nothing. He can only be as good as his work and, within reason, it doesn't matter what he looks like. There must be no attempts to register lurchers with the Kennel Club, to write 'breed standards' or any other adjuncts of the dog-show world. There is only one 'standard' for the lurcher and that is his work.

This is not a definitive book about "The Lurcher", nor is it intended to be. It is a tribute to the lurchers, longdogs and whippets that have given me so much fun, sport, excitement and exercise.

Ted Walsh
January, 1984.

Acknowledgements and Thanks

Anyone setting out to write a book about lurchers (and, so far as I know, this is the first one to be written entirely about them) must either know everything about lurchers or must ask some questions. Even Methuselah did not live long enough to have tried all the permutations of lurcher breeding; I am not yet of Biblical age and I certainly don't make any claim to know everything about lurchers so my first thanks to the many people who answered my questions so willingly. That I may not always have agreed with the answer I was given is neither here nor there.

I am most grateful to Mr Vesey-Fitzgerald for permission to quote from *It's My Delight*; to Mr Drabble and Michael Joseph Ltd for permission to quote from *Of Pedigree Unknown*, and to Mr Ian Niall and the Editor of *Country Life* for permission to quote from 'A Countryman's Notes'. And I acknowledge my debt to the shades of Sir Samuel Baker, William Scrope, Esq., and Charles St John, Esq.; when we eventually meet on the Elysian Coursing Fields I shall make a point of telling them how much pleasure their books give me.

I owe particular thanks to the people who have put pen to paper for me; George Smith who wrote the very full description of lamping in Chapter Six, Mr M. H. Salmon from Minnesota who sent me some very good photos. John and Gill Moore who did all the hard work in Australia and from that country Messrs Frank Roberts, S. T. Goodwin, Greg Kilgour, Robert Burke, Ray Hewitt, Bob Deering, Anthony O'Shea, W. R. Bloomfield, and Chas Venables. They not only answered what may have seemed to them to be silly questions but actually came back for more; "Good hunting, all of you". On the writing side I must also thank the staff of that marvellous institution, the Bodleian Library at Oxford.

Lurchers vary so much that I could not have considered producing this book without copious pictures. My special thanks to that Prince of sporting photographers, Jim Meads. Not only did he take more than half the photos in the book but he travelled many miles to do so and kept me up to the mark when my attention wandered. My thanks also to the many people who willingly allowed their dogs to be photographed; I hope they are not disappointed with the results. Others who very kindly provided me with photographs include Major Compton, Mick Giles, Brigadier Dunn, Maria Pether, R. & H. Chapman of Buckingham, Susan Waters, the Editor of *The Field* (and through him my thanks to Dereck Tilley, Harris Duff, Manfred Danegger, John Marchington, G. Kenneth Whitehead, E. P. Gee, C. H. Stockley, and John Tarlton), M. H. Salmon, Chas Venables, and Associated Newspapers Group. The remaining photos were taken by myself.

Finally my thanks to Anne Boyes who did the bulk of the typing for me despite her typing machine seizing up and me looking over her shoulder.

June 1977   E. G. Walsh

CHAPTER ONE

The Origins of the Lurcher

Of all the domesticated dogs that we know today the only two that probably look the same now as they did 1,000 years ago are the saluki and the lurcher. The saluki has not changed because he lived an isolated life in the deserts of Arabia, he was treated by his Arab masters as part of the family, life for man, dog and camel was hard, weaklings did not survive, and no cross-breeding was permitted. The lurcher has not changed because he was and is a dog bred purely for work; and that work has not changed throughout the centuries. The only difference is that until the late 1600s he was, at least in print, known as a greyhound.

The longdog, the sight hound, the running dog, or whatever one likes to call him, probably originated in the Middle East. Drawings and rock carvings more than 4,000 years old show a coursing dog not unlike a saluki with pricked ears. Isolated in the Arabian deserts the saluki has remained a pure breed but as the longdog spread out from the Middle East the type altered to suit local conditions, and in time there evolved the Afghan hound, the Borzoi, the Pharoah hound, the Ibizan hound, the Italian greyhound, the whippet, the Irish wolfhound, the Scotch deerhound, and, fastest of all, the greyhound. But the greyhound as we know him today at Altcar or Buckworth, at White City or Wembley, is a modern breed; only 130 years ago 'Stonehenge', himself a member of the newly formed National Coursing Club committee, was able to describe four distinct types of greyhound.

It is not certain when the greyhound first arrived in the British Isles but the probability is that the various waves of Celtic invasions, particularly the Belgae, brought their dogs with them. The first recorded mention of the word is c. 1,000 AD and in 1016, King Canute, "a Dane and a King of this Realme" granted at a Parliament held at Winchester the Carta de Foresta, the Forests Charter, which laid down how the Royal forests should be administered. Amongst many sections are several

"THE LURCHER", by Reinagle. According to contemporary writers (c. 1800) this dog would have been a rough, greyhound-sheepdog cross

"BETH", a greyhound-"Smithfield" cross

Jim Meads

"TESS"; a good example of a "Smithfield"

"BIFF"; a lurcher-greyhound cross. A very good working dog and a winner at several shows

dealing with dogs and section 31 starts with the words *"Nullus mediocris habebit nec custodiet canes quos Angli* greyhounds *appellant"*.

In Mediaeval England a forest was a definite tract of land within which a particular body of law was enforced, having for its object the preservation of certain animals; these were the game animals, the red deer, the fallow deer, the roe and the wild boar. In certain areas the hare was included. The forests were not necessarily woodlands but included waste lands, commons, heaths and both open and dense woodlands. Most of the forests were the property of the Crown but from time to time the King transferred some of them to his subjects as gifts, bribes or payment for services. In addition to the forests there were chases, where the game animals were preserved (but where the whole body of Forest laws might not be in force), and parks, which were districts enclosed by paling or other fencing; as a general rule they were enclosures for deer. And there were also the warrens: 'warren' meant either a right to hunt and take game animals on a particular piece of land, or it could mean the land over which such a right extended. Whilst the charter of warren did not preserve the beasts themselves it did reserve the right of hunting them to certain individuals. William II may well have "loved the red deer as a father", but whilst the King found sport in the Royal forests, the main reason for their existence – and, indeed for the parks, chases and right of warren of the nobles, both lay and clerical, and lesser landowners – was the prosaic one of sheer survival.

All, however rich, had of necessity to live on the produce of their land. The King had to feed and maintain his immediate Court, which could number from two to three hundred persons, besides the hangers-on who took what they could until discovered by the Stewards and Marshals. There was no distribution system for food and the only way of victualling this number was for the Court to move from Royal forest to Royal forest, pausing for a week or two to eat what the huntsmen had gathered, and then moving on.

This progress could be varied by the King billeting himself and his Court on his subjects. A vivid description of what this could mean to a host who was not well endowed is given in the early chapters of Conan Doyle's *Sir Nigel*; the Lord Chandos' message that the King (Edward III) intends to stop at Tilford Manor sends Nigel Loring into Guildford with what remains of the family treasure to raise enough money to provide for their visitors. Mercifully, most of the Court is sent on to stay at Guildford Castle and the visit ends the family's immediate financial

"LARCH"; a deerhound–lurcher cross. A winner at several shows

"BIMBO"; a "Norfolk" lurcher

Jim Meads

"MIZ"; a sixth-generation lurcher

"SADIE"; a small whippet-"Norfolk" cross

Jim Meads

troubles; but it was an ever-present threat that could bankrupt all but the very rich. In one year of his reign, Edward III moved 75 times, i.e., about three times a fortnight.

In addition to these constant moves for fresh food, game would be requisitioned from other forests not on the immediate route or which the Court seldom visited; most of this game would be salted before despatch. In 1260, Henry III's huntsmen were sent to a forest on the Welsh Marches to take 80 swine, and 30 hinds; and in 1267 the order was for 60 does, 16 hinds and 60 wild boar. As well as game, there was an income from the proceeds and fines of the forest courts, from rents, and from the sale of timber and woodland produce; and gifts and licences to hunt all accrued to the Royal coffers or the Royal prestige.

In general, this right to hunt in forest, chase, park or warren extended from the King, through the nobles, lay and clerical, to knights and landed men but no lower in the social scale. The public, the common people, had rights of hunting on what land there was left outside the forests, chases and parks (it is estimated that the Norman Kings appropriated over a third of the country as forests), but such land was very limited indeed and it held few game animals. For those who wished to augment their frugal and monotonous diet – a diet that could easily verge on starvation – the only alternative was to poach the King's or nobles' game or the warrener's hares.

Today, if one wants to find out about cases of poaching one looks at the police court reports in the local paper; from the 11th to the 14th century the equivalent were the forest eyre, the courts called into being by the King, appointing justices to hear the pleas of the forests in certain counties at a time. No matter what court's or year's report one looks at the poaching of deer with longdogs appears over and over again. Circumstances vary; sometimes bows and arrows or crossbows are used as well, sometimes it is one dog, sometimes two or three; on occasion it is as many as 14; but though there are some seven or eight hunting dogs differentiated, the word always used for the coursing dog, whatever he looked like, is *leporarius* which is translated as *greyhound*.

In *The Master of Game*, the oldest book on the chase in the English language, written by the Duke of York between 1406 and 1413, the greyhound is described as "a kind of hound there be few which have not seen some". The author goes on to say: "The good greyhound should be of the middle size, neither too big nor too little, and then he is good for all beasts. If he were too big he is nought for small beasts, and if he were too little he were nought for the great beasts. Nevertheless whoso can maintain both, it is good that he have both of the great and the small, and of the middle size." The accompanying illustrations show both smooth-coated and rough-coated greyhounds.

In 1570, Dr Caius, the founder of the college of that name at Cambridge, writing to Dr Gesner, the Swiss naturalist, describes the *leporarius* as "so called from its speed. Its value and its use are found in the hunting of hares. Although in catching fallow deer, stags, roebucks, foxes, etc., they excell in strength and traditional speed, their excellence varies with temperament and plumpness or slimness of body. For there is a thin sort comprising a larger and a smaller variety; and some have smooth hair". Dr Caius' description could apply to half the dogs illustrated in this book.

It is probable that there were pure-bred greyhounds in some numbers in Mediaeval England but they would have belonged to the wealthy and ennobled; and even for them the problem would have been one of finding the right dog to put a bitch to. There was a very considerable network of roads, the main part of which dated from the Roman Empire; but the maintenance of them was, in theory, the responsibility of landed proprietors and the Church. As a result, the repair of roads depended upon chance, goodwill, opportunity or the piety of those to whom the adjoining land belonged and the highway varied from a firm, dry surface to a bottomless swamp. Travel was often not easy for those who had to travel; for the man who wanted to breed a litter of 'greyhounds' there was no question of sending the bitch away to such-and-such a dog at a distance. Breeding was done from what was available locally and if the dog was not a pure-bred greyhound then what matter so long as his stock could gallop and kill? So Dr Caius' description, "a larger and a smaller variety; and some have smooth hair" is understandable for the majority of 'greyhounds' of his time.

CHAPTER TWO

The Lurcher in Print

The first mention of the lurcher in print is in 1668 by, surprisingly enough, the then Dean of Ripon, one John Wilkins, D.D. He wrote, for the Royal Society of which he was a Fellow, "An Essay towards a Real Charachter and a Philosophical Language; Tables concerning the Species of the Natural Bodies . . .". Chapter Five, of the Second Part, is on animals, and he gives a table of "The Rapacious Beasts of the Dog Kind". A clearly-laid-out chart shows the various dogs with their methods of hunting. Of those that hunt by "Swiftness and Running After", he gives "Greater Beasts: Greyhounds. Lesser Beasts: Lurchers". From this it would appear that by the late 17th century the lurcher was a recognised type and was smaller than the greyhound.

As a fascinating sideline, the Dean finishes this Chapter with a learned discussion on the animals that were taken into the Ark, the size of stall needed for each one and the amount of food required for the year that the Ark would be afloat. He details the carnivorous animals that were to embark, 40 in all or 20 pairs, and estimates that they would have needed five sheep per day. Room therefore had to be provided in the lower deck for the year's rations of 1,825 sheep, and hay to feed the sheep on a decreasing scale. There are also detailed calculations on the amount of hay required to feed the grazing animals on board. He shows that there were two foxes in the Ark and two dogs; but, as he does not say what sort of dogs they were, the theory that no foxes could have disembarked on Mount Ararat because of the two foxhounds and two terriers on board may not be correct!

The next mention of the lurcher in print is an Act of Charles II in 1670 which authorises gamekeepers to seize "Gunns, Bowes, Greyhounds, setting dogs, Lurchers or other dogs to kill hares". In 1675 the *London Gazette* contains an advertisement "Lost, a pied dog, somewhat shaped like a lurcher".

From here on, references to lurchers become more numerous and as books start to be written about field sports and sporting dogs so the lurcher is sometimes included, though not all the things written about him are complimentary: "A kind of dog much like a mungril greyhound". In 1760, Thomas Fairfax in *The Compleat Sportsman* says of the lurcher, "Lurchers is a kind of hunting dog much like a mongrel greyhound, with prickt cars, a shagged coat, and generally of a yellowish-white colour; they are very swift runners so that if they get between the burrows and the conies they seldom miss; and this is their common practice in hunting; yet they use other subtilties, as the tumbler does, some of them bringing in their game and those are the best. It is also observable that a lurcher will run down a hare at a stretch." His reference to the tumbler is interesting; the dictionary meaning is: "A dog like a small greyhound, formerly used to catch rabbits; a lurcher. So called from its action in taking its quarry." There are written references to the tumbler a good deal earlier than to the lurcher: "Tumblers, houndes that can goo an huntynge by then selfe; brynge home theyr praye" (1519). Fairfax gives a detailed description of how the tumbler catches his prey (which always appears to be rabbits): "So called because in hunting they turn and tumble, winding their bodies about circularly . . .".

William Taplin, writing in 1804 of the lurcher, tells us that: "The dog passing under this denomination is supposed to have been originally produced from a cross between the shepherd's dog and the greyhound, which, from breeding in-and-in with the latter has so refined upon the first change that very little of the shepherd's dog seems to be retained in the stock, its patience docility and fidelity excepted. The lurcher, if thus bred without any further collateral crosses, is about three-fourths the height and size of the full-grown greyhound and of a yellowish or sandy-red colour, rough and wirey-haired with ears naturally erect but drooping a little at the point; of great speed, courage and sagacity and fidelity; by which pedigree and appearance they are neither more nor less than a bastard greyhound, with some additional qualifications but without their beauty. These dogs, little calculated for the sports of the great, are seldom seen or known in the metropolis or its environs; but on the contrary are the established favourites of the holders of small farms, with many of whom they officiate in the capacity of the shepherd's dog, though they have speed and cunning enough to turn up a rabbit, or occasionally (when opportunity offers) to trip up a leveret, half or three-quarters grown, without the owner possessing either licence

Jim Meads

"PEGGY"; Lambourn Reserve Champion 1973 and Champion, 1976

"REBEL"; Badby Champion, 1976. He is part Irish Wolfhound

Jim Meads

Jim Meads

"FANNY"; second in the Rough Bitch Class at Lambourn, 1976, and winner of the large lurcher racing (see page 141)

"BERTIE"; a gypsy-bred dog. Lambourn Champion in 1972 and again in 1977!

Jim Meads

or certificate. . . . Prevented by nature from every chance of dependant society with the great, he calmly resigns himself to the fate so evidently prepared for him and so truly consonant with the predominant propensities of his disposition. Hence we find him almost invariably in the possession of, and in constant association with, poachers of the most unprincipled and abandoned description; for whose services of nocturnal depredation of various kinds they seem in every way inherently qualified. . . . They equal, if not exceed any other dog in sagacity. . . . Some of the best bred lurchers are but little inferior in speed to many well-formed greyhounds; rabbits they kill to a certainty if they are any distance from home; and when a rabbit is started not far from a warren the dog invariably runs for the burrow; in doing which he seldom fails in his attempt but generally secures his prey. His qualifications, natural and acquired, go still somewhat farther; in nocturnal excursions he progressively becomes proficient and will easily and readily pull down a fallow deer so soon as the signal is given for pursuit; which done, he will explore the way to his master and conduct him to the game subdued, wherever he may have left it. To the success of poaching they are in every way instrumental and, more particularly, in the almost incredible destruction of hares; for when the nets are fixed at the gates and the wires at the meuses, they are despatched by a single word of command to scour the field, paddock or plantation, which, by their running mute is effected so silently that a harvest is soon obtained in a plentiful county with little fear of detection." It is somewhat heavy going in parts but this was his style of writing; and underneath all the words it is obvious that William Taplin knew a good lurcher and was not unacquainted with poachers.

A writer whose description of the lurcher we can rely on is 'Stonehenge', or Mr J. H. Walsh, a sometime Editor of *The Field,* a writer of many books on dogs, who, when the National Coursing Club was established in 1858, was one of the members of the first Committee of three. Writing in the 1850s he says: "He [the lurcher] partakes of the points of the greyhound in shape, combined with the stouter frame, larger ears and rougher coat of the sheepdog, but varying according to the breed of each employed in producing the cross. . . . When the lurcher is bred from the rough Scotch greyhound [not the deerhound] and the collie, or even the English sheepdog, he is a very handsome dog, and even more so than either of his progenitors when pure. He is also a most destructive animal, showing speed, sagacity and nose in an extraordinary degree, from which causes the breed is discouraged, as he would exterminate all the furred game in a very short time. A poacher possessing such an animal seldom keeps him long, every keeper being on the look-out, and putting a charge of shot into him at the first opportunity; and as these must occur of necessity, the poacher does not often attempt to rear the dog which would suit him best, but contents himself with one that will not so much attract the notice of those who watch him."

Coming to modern times, there are several well-known countrymen-authors who have written on the lurcher. Brian Vesey-Fitzgerald, one-time Editor of *The Field,* one-time Secretary of the Gypsy Lore Society, author of many books on natural history, on Gypsies, on poachers and on dogs of all sorts, says: "The outcast of dogdom, not recognised as a breed by any kennel club or society, and frowned on by dog-lovers and sportsmen. But the true lurcher is a true breed all the same, breeding true to type and very carefully bred by certain Gypsy families. Originally a cross between the greyhound and the Bedlington terrier. A dog of great hardiness, speed and wisdom, which can be trained to do all manner of almost incredible feats. A true lurcher should not exceed 24 inches in height and should weigh about 50 lbs." And again: "All Gypsy dogs are called lurchers. The word has come to mean any longdog of doubtful appearance and savage temperament and obviously doubtful pedigree. But the true lurcher is a cross between the greyhound and the collie and nothing else. This is a cross you breed true and I know several Gypsy families that do breed true lurchers from lurcher parents."

Next, Phil Drabble, the naturalist and broadcaster: "A lurcher is not a mongrel but a deliberate cross-bred, the traditional breed of Gypsies who want dogs more for pot-boilers than for sport. Pure greyhound blood is diluted with a breed to make up for some of the deficiencies of greyhounds, and the outcross used varies with district or personal choice. Traditionally, sheepdog is the cross. The progeny have longer coats more stamina and more brains than greyhounds. In Norfolk, a notable area for lurchers, they still use 'Smithfields' or Norfolk lurchers. The old cattle dogs that used to herd the fat cattle to Smithfield were lanky, long-coated dogs about the build of Bobtails or Old English sheepdogs. They had brains, stamina and weather-resistant coats. The first cross was not fast enough to come up with a hare that had a good start, though they might be able to wear her down if the course could be prolonged without her taking cover. Ideally the lurcher should have three-quarter

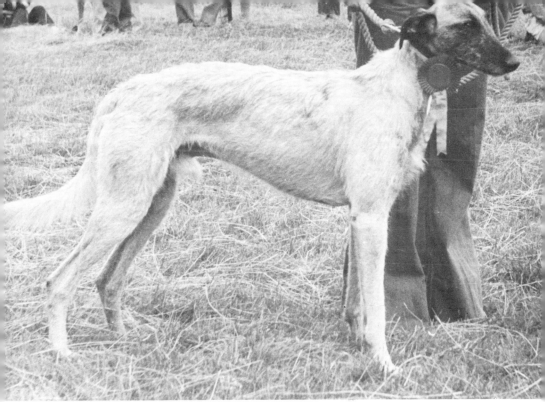

"PADGE"; litter brother to "TI" and "TARN" on page 151 and Scottish Champion 1977

"DON GIOVANNI"; a Lambourn winner and a great hunter and jumper

Peter Goulding

"ZIP"; a very nice small lurcher

"WAFER"; another nice small lurcher

Peter Goulding

greyhound blood, procured either by mating a half-bred dog back to a thoroughbred bitch, or, less often, by mating two lurchers, each three-quarter bred. This was the less common practice. Other outcrosses besides Smithfield sheepdogs were popular. Deerhound makes a wonderful cross, having good nose, superb eyesight, tough coat and temperament to match. Some people prefer greyhound-retriever cross. They used flat or curly coated retrievers, as labradors would be too slow. On the Welsh borders, working sheepdog was popular though not really big enough; Bedlington was sometimes used, though better as a cross with whippets for rabbiting; and since saluki has become popular for legitimate coursing, it is not uncommon to find saluki lurchers. My objection to salukis, which are supposed to have immense staying power, is that I have seen several that only gallop instead of really try. Anything can stay if it is never really extended."

And finally, Ian Niall, countryman, author and regular contributor to *Country Life*: "Time seems to cloak things that were once frowned upon or denounced, with a certain degree of respectability. Keeping a lurcher is one of them, judging from a television programme I saw not long ago. The word lurcher is synonymous with thieving. Once upon a time keeping one was like a housebreaker walking about with a jemmy in his hand. Whether he was in full employment or not, the cottager who owned this highly intelligent cross between a collie and a greyhound, came under suspicion of poaching."

So, whatever the relationship between the greyhound and the lurcher in the Middle Ages – for my part I am satisfied that the word greyhound covered the pure-bred, such as they were, and the lurcher – for the last 300 years there is general agreement that the lurcher is a cross between the greyhound and the sheepdog with an occasional other breed being used. Let us see what the present-day market place says; from one edition of *Exchange and Mart* we find offered for sale:

borzoi × greyhound bitch
deerhound greyhound × greyhound
greyhound × Bedlington whippet
whippet × greyhound collie
Manchester terrier × whippet
saluki greyhound × deerhound greyhound
surly greyhound × Norfolk [was this a misprint or was it a true
    description?]

Bedlington × greyhound
collie × greyhound
whippet greyhound × deerhound greyhound
Bedlington × whippet
greyhound whippet saluki × greyhound whippet Bedlington
greyhound collie × greyhound whippet
staghound greyhound × greyhound whippet

You can pay your money – and it can be big money – and you take your choice.

If the pure-bred greyhound is so fast that the Rules of Coursing say the hare must be given 80 to 100 yards start before the two dogs are slipped, it may well be asked why lurchers are bred at all. There are several reasons, the main one being that the lurcher's task has always been the finding and catching of game for the pot, whereas the greyhound has been used, and is used, for the sport of coursing.

Coursing is the pursuit of game with dogs that hunt by sight; but for a very long time the word has been used for a competition between two dogs to see which is the best, the hare being used to provide that competition. Eighteen hundred years ago Arrian wrote that "The true sportsman does not take out his dogs to destroy the hares, but for the sake of the course, and the contest between the dogs and the hares, and is glad if the hare escapes." Coursing as a national sport can be dated from the time of Elizabeth I when the then Duke of Norfolk drew up the rules which, in essence, still govern the sport. The contest is to decide which is the better of two dogs; the winner of a course is the dog that shows the most speed, agility and courage. "The Judge shall decide all courses on the one uniform principle that the dog that scores the greatest number of points during the continuance of the course is to be declared the winner." From the rules written by the Duke of Norfolk 400 years ago to the present-day rules of the National Coursing Club – or of the Whippet or Deerhound or Saluki Coursing Clubs – there is no mention of the dog that kills the hare being the winner of a course. The dogs strive to catch the hare, otherwise there would be no coursing. But the kill, when it does happen, is usually incidental to the course and seldom makes any difference to the result.

To start with, the hare is given 80–100 yards start before the two dogs are slipped; this distance is enough to take the edge off the greyhounds' tremendous speed. Then, the greyhounds' storming rush on the line of the hare is foiled, time and time again, by the hare's ability to stop and turn from full gallop; to turn back on its tracks and into full gallop again whilst the dogs are slowing down, turning and accelerating on the hare's new line. In seven cases out of ten the hare will reach cover or a hole in the hedge surrounding the coursing ground without the greyhounds touching it and the course is over. All of which makes a wonderful spectacle, but it is not much good to the man who wants something for supper.

What the hungry man wants is a hare in the inside pocket of his jacket. Throughout the history of coursing the word 'lurch' or 'lurching' has been used as a term of abuse for a greyhound. One of the dictionary meanings of lurch is to 'run cunning', to cut corners, to anticipate the turns of the hare or to wait on the other dog and pick up the hare as he turns it. And this is exactly what a good lurcher will do. The mixture of sheepdog blood with the greyhound produces a dog that is slightly slower and a dog that has more brains than the pure greyhound. By not galloping quite so fast, he is not thrown out so far when the hare turns; and by anticipating the hare's turn, or waiting on the other dog if two are being used, he picks up the hare when the greyhound would be galloping straight on. And, more to the point, the lurcher's work is not just catching hares – or rabbits – by straightforward coursing. There are many other methods of catching game, purse-netting, gate-netting, long-netting, catching a pheasant as it flies from a hedge; all of which the greyhound × sheepdog lurcher can and will do, the sheepdog's herding instinct being made use of in driving game to nets.

CHAPTER THREE

The Gypsy's Lurcher

To most people who know nothing about lurchers, the word is synonymous with Gypsies. "Lurchers? Oh, you mean those Gypsy dogs", is the usual reaction. In fact, it would probably be nearer the truth to say that about 50% of Gyspy families kept lurchers and much less than this proportion bred them. A lurcher is not easily hidden, its type is unmistakeable, its use is obvious. The sight of a lurcher alerted every gamekeeper and country policeman until the family concerned had moved out of the parish. Many Gypsy families preferred a terrier and ferret and a pocketful of nets; some families kept no dogs at all or, at the most, a guard dog of sorts. But it is probably true to say that the few Gypsy families that did breed their own dogs, that kept their own blood-line and trained their own dogs, brought their dogs to a state of training that would have been beyond the abilities of most gundogs or police dogs.

To the outside enquirer – the *gajo* – the problem is to find out any answer that relates to the question. I make no claims to any special knowledge of Gypsies; and I have seen very few Gyspy dogs working and then only a bit of rabbiting or the equivalent. When questioned by a stranger, travellers, more often than not, give him the answer they think he wants. Like any other primitive people with a long history of oppression they are masters of the evasive answer. There may be some money in telling the man what he wants but it is more than likely that he is a bureaucrat preparing some form of greater or lesser oppression so there is no point in telling him the truth. When one looks at the history of the Gypsies, in this and in other countries, the attitude is understandable.

I have found that travelling people are more ready to talk if one has a longdog at one's heels when approaching them. If there is going to be any rapport at all, the uproar between the camp dogs and one's own is usually sufficient to break the ice. From then on it is played by ear, and

Six Open-lot wagons and one Bow-top leaving Yarm Fair. The dog on the footboard of the leading van is having an easy journey

by ear it has to be. Many Gypsies are asked questions by many enquirers into their language and habits and the sight of a tape recorder can mean either a clamp down or a spate of talk from everyone at once that is impossible to sort out afterwards. The fact that one merely wishes to talk about dogs makes little difference at first.

I do not believe that the Gypsies brought the lurcher to Britain; the lurcher was here when the Gypsies arrived, probably at the beginning of the 16th century, though a Scottish Act of 1449 mentions "overliers and masterful beggars going about the country with horses, hunds and other goods" and there are earlier reports of Gypsies on the Continent keeping hunting dogs like the nobility.

The Gypsies probably came from India originally. They travelled across Asia Minor and through Europe, keeping their own culture and language and assimilating what was useful from each country as they passed through. Persecution followed them wherever they went, varying from 'moving-on', fines and imprisonment to exportation and execution. This persecution did not cease with the Dark Ages; nor did it cease with Hitler and the Nazis who exterminated some 300,000 Gypsies in Germany and Eastern Europe. Persecution and harrasment persists to this day in Britain, kept alive by an insular fear of the unknown and a bureaucratic insistence on everyone and everything being classified and pigeon-holed.

When the Gypsies first came to Britain in about 1500 the country was sparsely populated. There were the forests, commons, moors, heaths and deserted tracks where travelling people could hide when oppression became too heavy. They had many trades; for cash they told fortunes, mended tin and copper-ware, traded in horses, made brooms and baskets, worked at piece-work on the land and, later, when prize fighting was the rage, fought with success and distinction. For food they bought or bartered what they had to buy and for the rest they lived off the land and many families kept dogs to help them do so. For anyone who has seen a tribe of Indian Gypsies on the move it is not difficult to imagine Gypsies in Britain in the Middle Ages. A family of Sansyas or Banjaras is not easily forgotten; a motley company of men, women and children travelling by side roads and unfrequented tracks; their possessions, baggage, tents, tent rods, pots and pans, chickens and babies, carried on buffaloes, donkeys or pack ponies; and an uncounted rabble of rough, hard-bitten longdogs of all sizes trotting along in the dust. Arriving at their night's camping place, the men would go off with nets

A Reading wagon (probably Dunton-built), in Northamptonshire. This is the popular view of the lurcher but, other considerations apart, this way of life has been made hazardous by modern traffic and road surfaces

and dogs to see what they could catch for the evening meal; fox, hare, jackal, iguana, ant-eater or even carrion, nothing came amiss to them.

Hunting by scent is noisy and takes time to accomplish; those who do not wish to advertise their presence use dogs that hunt by sight and hunt silently no matter what the temptation to 'open'. If there is time to set nets they will do so; driving game into nets is often simpler and more certain than straightforward coursing. The various acts against keeping 'greyhounds' would have applied to the newly arrived Gypsy as to the English countryman in the 16th century but with the amount of legislation already against him the Gypsy would not have paid much notice to a law that forbade him to keep a dog that provided him with his meat. The families that kept longdogs needed dogs that were intelligent and trainable as well as fast and strong. In times when each man had to safeguard his property as best he could, dogs were allowed to roam loose at night; the Gypsy who had a bitch in season would have marked down an appropriate mate – be it greyhound or sheepdog or whatever he required – and to the man who was used to moving quietly at night a stolen service presented few problems. Thus those Gypsy families who kept longdogs would have evolved the breed they wanted but the type would not have differed much: medium-sized, tough hunting dogs, probably with rough coats, that bred more or less true to type, well capable of filling the cooking pot and, like their masters, distrustful of strangers and capable of living rough in all weathers.

There was little change in the life-style of the Gypsy until the late 1800s when the living-van, or wagon, started to be built, developed from the vans used by travelling showmen. These vans became the mobile homes of most Gypsy families, supplemented with flat carts, traps, etc., for faster travel, excursions from the camping ground, or carrying extra accommodation. When the time came to move from the camp site or pitch, the van would be loaded, chickens or bantams caught up, and all made secure inside. Horses, brought in from grazing before first light – for they were probably feeding in someone's field without permission – were dressed and backed into shafts. Dogs were put in vans or tied to rear axles, the fire was doused and the vans started from their places; a tricky job if the ground was soft or the horses not up to their job or weight. The procession would make its way to the next pitch, with a stop or two during the day for a drink or a gossip. Dogs might or might not be loosed during the day; daylight poaching, though carrying lesser penalties than night poaching, was more precarious unless the ground was well known. In any case they would not go far from the road; any serious work would be done when all was settled at night. In theory, it was the perfect life for keeping a dog fit; fairly fast walking during the day and a gallop in the evening to keep the wind clear. But I have noticed that a lot of the travellers' dogs that I have come across – even those, now, from static sites – suffer from what used to be called kennel lameness: rheumatism, from wet and cold lying. An occupational hazard as are the inevitable tears from barbed wire.

In the post-war years of the late 1940s and early 1950s another major change came in the lives of many Gypsies and three things brought this change about. They were the need of the static, affluent and ever-enlarging population to rid itself of vast and increasing amounts of scrap, much of it scrap metal (Gypsies had been in the metal trade for many years when the need had not been so great); housing, pavements, street lighting, road making and widening, and all the horrors of 'civilisation' spreading over the countryside; and the changeover on the part of the Gypsies from horse-drawn vans to lorries and larger, shinier, towed caravans. The travelling men's trade of scrap dealing and tarmac-ing meant that they had to live nearer centres of population; the changeover to motor transport meant remaining on hard roads and using hard-cored lay-byes; and the greater cash flow meant that for many of them the meat once caught by their dogs could be bought in the butcher's shop. The medium sized, highly intelligent, rough-coated dogs, greyhound and sheepdog or deerhound and sheepdog-bred were replaced by larger, faster dogs, many of them a deerhound to greyhound cross, with less and less of the collie. Apart from some poaching the dogs are used for Sunday coursing; at times large sums of money change hands and, until recently, hares were often left where they had been killed. Inflation has caught up with coursing as with every other activity and with hares fetching £1.50 each at the time of writing, they are, like the fox's pelt, not lightly left to rot. There are still some Gypsies who live in vans; some from no wish to change, some from necessity. But more and more are settling on permanent sites as they become available, unenticing as many of such places are. The old way of life is no longer there; no one wants wooden clothes pegs, no one wants kettles mended, brooms and baskets are made of plastic and the few fortune tellers have their static pitches. There is still a certain amount of dealing in horseflesh but that is another story. The Gypsy's dog has changed with the Gypsy's way of life, and not always for the better.

CHAPTER FOUR

The Breeding of Lurchers

Most lurchers are bred from lurchers; it is as simple as that. Their parents on both sides are lurchers and the original crosses may be so far back that no one knows what they were. 'Norfolk' is the name given to many of them; medium sized, usually rough-coated, rather long in the back and often with hare feet. Provided that there is an injection of greyhound blood every third or fourth generation and litters are rigorously culled to remove any that are too small or show any sign of weakness, the person who merely wants to buy and is not interested in breeding his own need look no further; the difficulty lies in finding a strain that has been carefully bred. Many litters are bred from parents whose own breeding includes half a dozen different types; every puppy is reared and sold, a fictitious 'pedigree' is added and the fact that the great-grand-parents range from deerhound to Italian greyhound is not passed on – often because the breeder himself does not know it. Small wonder that such dogs produce very odd offspring. As in painting in oils or making a sauce, so in breeding lurchers: the fewer basic ingredients the better the result. Too many colours stirred up on the palette produce a grey-brown sludge; too many herbs and spices in the sauce and it eventually tastes of nothing; too many breeds of dog in a lurcher's pedigree and the breeder is lucky if one out of a litter is worth rearing.

Before considering the various breeds that are used to produce lurchers and longdogs today we should be clear about what is needed. The lurcher must have speed, stamina, brains, courage, nose, soundness and a weather-proof coat. The speed need not be quite that of the greyhound; indeed, it is the pure speed that tires out the greyhound so quickly. Stamina is essential to the dog that has to run down his game and repeat the exercise as soon as he has got his tongue in again. Without intelligence the lurcher cannot be trained in obedience; he must have courage to face thorn hedges, wire, rough going and water; he must have

Coursing greyhounds; "EIGHTSOME REEL" (above) and "WHISTLING GLORY" (below)

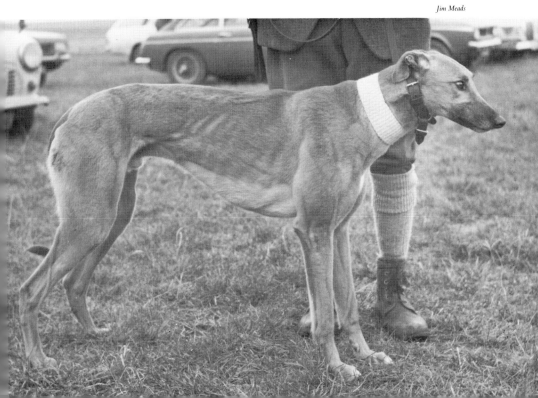

sufficient nose to follow up and retrieve wounded game. Soundness is an obvious requirement as, like any other working dog, he is useless if he breaks down. And a thin, single coat would prevent him working in all weathers. On this matter of coat, there are many lurcher-owners who say that a rough coat is necessary to avoid too much damage to the dog from wire etc. I do not wholly agree with this argument. The dog must have a weather-proof coat but in my opinion damage is more a question of the individual dog. I have had dogs that have worked through to old age without a mark on them; I had others that have looked like exhibitions at a show of needlework by the time they were three or four years old. One dog will cut himself going through an open farm gate, another will go flat out through pig netting laced with barbed wire without getting a scratch on him.

With all these attributes, the lurcher must not be too small or he will only be good for rabbiting; and he must not be too big or he will be ungainly, unable to turn with or pick up rabbits and hares. In my opinion a lurcher should be from 24 inches to 26 inches at the shoulder and from 45 lbs to 65 lbs when fit, though I know of many good dogs outside these limits.

Before discussing the various breeds I would enter one *caveat*; it is a point that is obvious to anyone brought up with working dogs but not so obvious to beginners. It is, to avoid, above all things, breeding from any dog from a show kennel or from a show background (I am here, of course, talking about the various pure breeds). Dogs that are bred for show are bred to an arbitrary standard laid down by a breed society. This standard will be written purely from the point of view of what the dog looks like; its height, colour and length of coat, colour of eyes and nose, length of ears, whether the ears are pricked or hang down, the curl or lack of curl in the tail, etc., etc. None of which has the slightest bearing on whether or not the dog can do the work for which it was originally bred. There are three exceptions to this rule: whippets, deerhounds and salukis, or, at least, some of them. There are some 150 whippets and somewhat fewer deerhounds and salukis, running at Whippet, Deerhound and Saluki Coursing Club meetings each season; quite a lot of these dogs also appear in the show ring and they are all pure-bred and registered at the Kennel Club. If one is using any of these three breeds then the kennels to go to are the ones that course their dogs. At the other end of the scale are the show greyhounds and the show collies; they are separate breeds from the coursing greyhounds and the working collies and can only be expected to trot up and down a strip of coir matting in a drill hall.

As we have seen the lurcher was, by general agreement, a cross between the greyhound and the sheepdog. It is a cross that will usually breed true for generations though from time to time fresh greyhound blood is required to keep up speed. Provided that care is taken not to breed from rubbish, this cross should satisfy the wants of most lurcher-owners and I do not see the point in using any other breed except for a touch of deerhound. There are now all sorts of dogs advertised for sale as lurchers and many of the dogs produced at lurcher shows are ones that I would pay rather than own. Some of them do have a look of what they are said to be, some could be almost any cross and a lot of them should have been put down at birth. Still, these extraordinary mixtures do appear and it is worth while looking at the breeds used and see what they have to offer the lurcher-owner.

Greyhound
The early greyhounds were of differing sizes and types of coat. As recently as the 1850s, 'Stonehenge' described four different dogs: The Newmarket dog, of great size, immense speed and a tolerable degree of stoutness; The Wiltshire dog, a small muscular, compact animal, more like a terrier, with untiring energy and great working powers . . . many of these little greyhounds not weighing more than 30 to 35 lbs; The Lancashire dog, bred exclusively for the plains of Altcar and Lytham and those of Lincolnshire and Cheshire. Here the dog had not only to be fast to his game but had also to be able to see it 100 yards away when up to his elbows in stubble. Much was therefore sacrificed to size and speed, even more so than at Newmarket; and as the Judge was often not able to follow a course on horseback, due to the soft ground and the dykes, the first part of the course was generally all he saw and so speed was more important than stoutness. Finally 'Stonehenge' mentions the Scotch dog which "can scarcely be considered a distinct variety" but which, elsewhere he calls the "rough or Scotch greyhound" and which other contemporary writers describe as rough coated, coming from the Lowlands of Scotland and owing its coat to the deerhound of the Highlands.

As travel became easier, particularly with the spread of railways, it became possible to take a bitch to a dog out of one's own district and to attend distant coursing meetings, taking dogs by train and cab, and with this movement of people and dogs the different types of greyhound gradually disappeared. There can still be a fairly large variation in size between one dog and another and, of course, between dogs and bitches; coursing dogs tend to be somewhat smaller and stockier than track dogs. But size, in itself, is not necessarily a recommendation; the over-large dog can take half a parish in which to turn whilst his smaller rival is tied to the hare's scut.

A criticism of the modern greyhound is that he can be short on stamina. The days are far gone when 'Stonehenge' talked about the daily exercise "increasing from fifteen to twenty-five miles a day", and "galloping at best pace a mile and a half to two miles"; when 'The Druid' described a course as "watching them work her [the hare] for nearly three miles on the hillside"; of Harding·Cox advocating "twelve miles at a long slinging trot behind a horse" saying that stamina was as important as pace. Nowadays the call is for speed and still more speed with courses lasting 50 to 60 seconds; the few printed reports merely tell of the lead to the hare and say almost nothing of subsequent work. Many running grounds are surrounded by wall, hedge or split chestnut fence with escapes for the hare; and on the track, a dog that can gallop for 700 yards is called a stayer!

Nevertheless, the greyhound is the fastest of all domesticated dogs and, to me, it is a joy just to look at a coursing greyhound walking, head held low, with that wonderful thrust from the stifle, muscles rippling and a whole expression of pent-up power.

The lurcher breeder must remember that there are good and bad greyhounds as there are good and bad of other breeds; he must particularly beware of using a dog that has any tendency to bad feet. The otherwise good greyhound with doubtful feet can, with proper care and attention, be trained to the minute to win a stake. But the lurcher must be able to pull out sound at a moment's notice and there are enough natural hazards in the lurcher's life without his owner having to contend with bad feet as well.

Sheepdogs and Cattle Dogs

With his speed coming from the greyhound, the other half of the true lurcher, the brains, stamina and weather-proof coat, comes from the sheepdog or cattle dog. These vary from the bobtail that the Sussex and Hampshire shepherds used on the heavy clay of the Weald and the chalk

Jim Meads

Coursing whippets; both dogs also appear, very successfully, in the show ring, the older dog on the left being a Champion with 10 C.C.s

A coursing deerhound

Mick Giles

Jim Meads

"BLUEY"; a very good example of a "Smithfield"

"TINA"; a litter sister to "BLUEY", above

Jim Meads

of the Downs to the aristocrat of the trials, the Border collie that can hold sheep with its 'eye'.

'Stonehenge' said of the dog of 100 years ago: "The English sheepdog has a sharp muzzle, medium-sized head, with small and piercing eye; a well-shaped body formed after the model of a strong, low greyhound, but clothed in thick and somewhat woolly hair which is particularly strong about the neck and chest. The tail is naturally long and bushy . . . in almost all sheepdogs there is a double dew claw on each hind leg. The legs and feet are strong and well-formed and stand road work well, and the untiring nature of the dog is very remarkable. Such is the true old English sheepdog, but a great proportion of those in actual use are crossed with the various sporting dogs such as the setter, which is very common, or the pointer or even the hound; and hence, we find the sheepdog as good in finding game as in his more regular duties, while a great many are used as regular poaching dogs by night and, in retired districts, by day." There has been little change in this description over the years. The important thing about the sheepdog from the lurcher breeder's point of view is that he is a working dog and is bred to do a job of work, not to look pretty. If he cannot pull his weight on the farm he is not kept. There are hundreds and hundreds of fast, strong, intelligent working dogs in almost every county and the present-day lurcher breeder merely has to look and he will find.

There is no need to go into details of the sheepdog, but in discussing lurchers it is of interest to consider the Smithfield, or drover's dog, since 'Smithfield' is a name often used in connection with lurchers. Cattle and sheep have not been driven to market 'on the hoof' for a great many years, but until the coming of the cattle truck there was no other means of getting them to town, be the town London or any other market town.

William Taplin (1804) says of the Drover's Dog or Cur: "This dog, though with some points of similitude, is both larger and more ferocious than the shepherd's dog to whom, in appearance he evidently seems akin, but stands considerably higher on his legs with a much more commanding aspect. In colour, the cur is of a black, brindled or a dingey-grizzled brown, having generally a white neck and some white about the belly, face and legs; sharp nose; ears half-pricked and the points pendulous; coat, mostly long, rough and matted, particularly about the haunches, giving him a ragged appearance to which his posterior nakedness greatly contributes, the most of this breed being whelped with a stump tail. This dog, from his formation and complicated appearance, was in its origin most probably engendered between the shepherd's dog and the lurcher with an intermediate cross of the mastiff or Dane. He is a serene and faithful follower of fortune with his employer and is seldom without the power and ability to render assistance in any little poaching excursion that may be occasionally entered into at certain seasons of the year." Another author, writing in 1851, says: "Closely allied to the shepherd's dog is the cur, or drover's dog. This useful animal is larger than the shepherd's dog, the hair is generally shorter, and the tail, even when not cut purposely, often appears as if it had been so. It seems to us that the drover's dog is, in reality, a cross between the shepherd's dog and some other race. These dogs are singularly quick and prompt in their actions and, as all who have watched them in the crowded, noisy, tumultuous assemblage of men and beasts in Smithfield must have observed, they are both courageous and intelligent."

At the turn of the century, a report headed "An Old-Time Drover" said, "Old Bill Thacker, of Lincolnshire, is nearing ninety years of age and is probably the only drover left of his class in the country. Scores and scores of times, sixty or seventy years ago, Thacker has passed through the eastern counties with large droves of cattle on his way to Norwich, London and other markets. The journey to London occupied four to five days and the drover, or 'topsman' as he was called, having got rid of his stock, would walk back into Lincolnshire and pick up another drove and return to London or some other centre again. A thirty, forty or even fifty miles' walk was nothing. It was no uncommon thing for a drover to have the care of five thousand to six thousand sheep or three hundred or four hundred head of cattle for ten to twelve days. The great objection of the drovers in those days was not the roads – difficult to negotiate as they were – but to the old toll-bars, which not only meant the payment of large sums as toll, but also caused irritating delays, the toll-keepers being described as not of the most amiable type."

I am sure, from evidence and from inference, that there never was a breed of drover's dog but there was certainly a type. Picture the roads and the conditions that the drover had to face, remembering that, where it would not take him too far out of his way he would use the side roads to avoid toll-gates. A week or more on the journey with no cover at night for beasts; roads with, at best, gravel surfaces and, at worst, bottomless mud. Wide verges where cattle could pick a bit of grass and often no hedges, particularly when crossing a heath or common where

Jim Meads

"QUEENIE"; a very shaggy dog but she has bred some good lurchers: see her grand-son, "WHISKEY", below

"WHISKEY"; a greyhound-"Smithfield" cross

Jim Meads

cattle could stray. The dogs had to have harsh, weather-proof coats, they had to have the courage to outface horned cattle and frightened cattle as well, they had to have the speed and strength to reach and turn beasts that did not intend to turn; and they had to have the intelligence to work sheep and cattle through towns with their many side streets, traffic and pedestrians. One might add that they needed an iron constitution; the drovers themselves had to be hard men and one only has to think of how market men treated beasts up to quite recently to imagine how beasts and dogs were treated 100 years ago.

Though cattle are no longer driven to market by road, the type of the drover's dog is by no means extinct; one sees them on farms and at markets. Only recently, when visiting a neighbouring farmer, I noticed that his large black and white collie was no longer in the yard and that there was a three-month-old puppy in its place. On my commenting, my host said that the dog had been "a bit sharp; he was inclined to bite the cows if they didn't move quick enough for him and it got so that the postman wouldn't get out of the van if the dog was around. He's gone up to near Daventry to a man who does a lot of cattle dealing; he'll make a good drover's dog". There is, I understand, a breed of sheepdog in Tasmania that are known as Smithfields and, certainly, photographs of dogs on sheep stations in New Zealand show many types that could well be drovers' dogs of 100 years ago.

Breeding from Scratch

If I had the time and money and the space to start again and breed lurchers from scratch, I would start with two unrelated sheepdog bitches. Exactly what they were would have to depend on what I could get; I would like one to be a Sussex bobtail but I would be very surprised if there are any of them left now. I would not necessarily look for trial blood but both dogs would have to be of working farm stock; if possible, one would be rough and the other smooth coated. They would not be too small and one of them should certainly be over 22 inches.

One of the collies would be put to a coursing greyhound and the other to a coursing deerhound. I would cull the resulting litters hard, aiming to finish up with one dog and two bitches in each family; let us call them the A's and B's for ease of reference. They would be worked from about 15 months to try and find any weakness and the bitches would be bred from at their second heat, the A's going to the B dog and vice versa. The resulting four families would again be culled hard and I would hope to finish up with one bitch from each mating. The two A's would be 2/8th greyhound, 2/8th deerhound and 4/8th collie and the B's would be 2/8th deerhound, 2/8th greyhound and 4/8th collie. To adjust these proportions, each bitch would be put to an unrelated, coursing, greyhound which would produce a third generation of 5/8th greyhound, 2/8th collie and 1/8th deerhound; this, in theory and usually in practice, is the perfect breeding for a lurcher. With unrelated sires, this third generation could be used for cross-breeding and no outside blood would be needed until speed started to drop; outcrosses to coursing greyhounds would then be made as needed.

CHAPTER FIVE

Some of the Other Breeds Involved

I have discussed the traditional lurcher, the greyhound-collie cross and its variations; deerhound-collie, greyhound and deerhound-cattle dog cross, and their permutations. In my introduction I suggested some definitions; that for the lurcher being a cross between a coursing dog (indigenous) and a working dog. The coursing dogs are the greyhound, the whippet and the deerhound; working dogs are a larger number and one must consider the main ones whether or not they are suitable for breeding lurchers. I am well aware that there are some very good working dogs about that are of, to say the least, unfashionable breeding; this is always the case with working dogs where the emphasis is on work and not on looks. There is also the problem that many dogs carry out their work in private and one has only the owners' word for their excellence.

Therefore, in discussing the breeds that are sometimes used for producing lurchers I have kept to those most often met with and have been selective with these. I repeat my definitions, that a lurcher is a cross between a coursing dog and a working dog, and a longdog is a cross between coursing dog and coursing dog. Anything else is a mongrel; and by coursing dog I mean the indigenous breeds, the greyhound, whippet and deerhound. The saluki is an immigrant. So, to some of the possibilities, in alphabetical order.

Afghan Hound
There is, apparently, little or no ancient history of the breed but the Afghan Hound is probably an offshoot of the Saluki that adapted itself to the very rough terrain and the harsh climate of Afghanistan, with a thicker coat and a shorter and more powerful loin. The 'Afghan greyhound' is mentioned by some of the soldiers and travellers that visited that country in the last century and one or two old photographs

of the Guides, ·at Mardan, showed a dog that could have been an Afghan.

The dogs were kept for hunting gazelle and leopard both in the river valleys and up in the hills over the most appallingly rough country and it is interesting that they were often hunted in packs rather than singly or in pairs. Those people who have soldiered on the North West Frontier will have seen Afghan hounds, together with the bigger, shaggier and even worse tempered Powindars, travelling with camel caravans as guard dogs. One soon learnt to give them a wide berth unless one was armed with a heavy stick. These native dogs did not, for obvious reasons, have the exaggerated coats that are seen in the show ring today; their tatty, ragged hair finished at the knee and hock, giving them the appearance of wearing plus-fours.

I have only come across two Afghan cross-breds and did not see either of them working. The only pure-bred Afghan that I have seen coursing a hare did not impress with its action, speed or turning ability though it was said to have performed very well on a race track. Though they are large and powerful dogs and are, in their native country, used for very rough hunting indeed, they are extremely independent-minded and have to be watched very carefully with farm stock. They have an up-and-down action rather than a flat gallop and I do not think they offer much to the longdog breeder in this country.

Alsatian

The Alsatian is probably the most intelligent of all pure-bred dogs, though anyone who has used a standard poodle as a gun dog would challenge this statement. The Alsatian has highly developed senses, an ability to think for himself, a nose that is the equal of any gun dog, tremendous strength and stamina, a weather-proof coat and quite a fair turn of speed for a heavy dog. The present-day show-bred Alsatian has such an exaggerated bend of stifle that he can do little more than run round a ring trailing his tail on the ground, but there are a lot of Alsatians that are not so deformed and if the right sort can be found, the cross with the greyhound can produce a very powerful dog, not very long in the leg but able to gallop and with guts enough to tackle anything.

Since Alsatians are often found as guard dogs on caravan sites, the greyhound-Alsatian cross is quite common with scrap-metal dealers and purveyors of gates and tarmac. If choosing a puppy from a litter of this breeding, the ones that have a strong look of the Alsatian when young should be discarded; they will probably grow up too cobby to be fast enough for coursing.

Bedlington

As with so many others of the terriers, the origins of the Bedlington are obscure, but there have been dogs of this type on the Borders for at least 200 years. Various families (Armstrongs, Faas, Charltons) kept such dogs for working with foxhounds and otter hounds and for badger, otter, fox, rat, rabbit and other vermin on their own. Once the type was established it was often used as a cross with Lakeland, Border and other terriers for working with foxhounds. The working Bedlington looks like a cross between a whippet and an otter hound with a grey, linty coat, an arched back, almost a Roman nose, all the guts in the world and, when roused, the Hell of a temper. They were never put into an earth together; having killed their fox, two Bedlingtons would set about killing each other. An owner of working Bedlingtons has to use tact at all times and avoid situations where jealousy can lead to fighting. The trouble is that they usually fight silently so one does not know anything has started, and they fight to the death.

The greyhound-Bedlington or whippet-Bedlington lurcher has been used for a great many years along the Border and also in Wales where many Gypsies bred this cross. The whippet-Bedlington is often too small to be a consistent hare dog but there are few better all-rounders than the greyhound-Bedlington; the one drawback can be lack of size but it is rare that a whole litter turns out small.

Borzoi

Of all the pure-bred coursing dogs that have been used for hunting dangerous game – dangerous in the sense of being able to retaliate against dogs of any size, i.e., deer, boar or wolf – the only one that has been used on any scale up to recent times is the Borzoi. It may be uncertain whether the Borzoi started as a saluki type in the Middle East or whether he evolved from indigenous dogs in Northern Asia because of the use to which he was put but, whatever his origins, there are records of coursing dogs in Russia in the 13th century and a 'standard' for the Borzoi was written as early as 1650 which differs little from the present-day standard in Russia.

In a land as vast as Russia, noblemen's estates were often very much

"PERAK"; an Alsatian cross

"JASPER"; an Alsatian cross

"SPRING"; a greyhound–Bedlington cross

"SOLO"; a Bedlington–whippet cross

bigger than an English county and many families had their own hunts: Borzois in their hundreds supported by one or more packs of foxhound type; armies of huntsmen and retainers on matched horses. There was great formality in their style of hunting. Wolves were driven from covert by foxhounds; leashes of Borzois, usually two dogs and a bitch, matched in colour, were slipped from horseback. As often as not the wolf would be captured and put into an iron cage on wheels, to be released for future hunting or for training young dogs. Although the dogs were trained to knock a wolf off balance and hold it until the huntsmen got up to them, there must have been many occasions when Borzois had to deal with a wolf on their own, an animal as heavy as themselves and probably a better fighter. There were – and perhaps still are – certain tribes in Eastern Russia that used eagles to slow down the wolf just as falcons were used in Arabia to slow down gazelle when hunting with salukis.

Borzois in England are very seldom used for work. I may have been unlucky in the only ones I have seen coursing, but when I was privileged to watch them at a combined meeting in Norfolk, three brace were on the card. The first pair fought in slips and had to be removed. The second pair went 30 yards out of slips and started to play; and of the third pair, red collar went ten yards, stopped and squatted down to empty himself whilst white collar galloped straight back to his owner.

But there are many genuine working Borzois in other parts of the world. They are extensively used in Russia for catching foxes – and sometimes wolves – for the fur trade as the skin is very much less damaged than would be the case with shooting. In America, Borzois, or more often cross-bred Borzoi and greyhound or deerhound, are still used for controlling coyotes that prey on sheep and calves; and there are Coursing Clubs for pure-bred dogs, hunting hare and jack-rabbit. In Australia, Borzoi blood was sometimes used in the mixture of greyhound and deerhound that produced the kangaroo hound.

The only Borzoi lurchers I have seen – or, more properly, longdogs – have been too tall and thin to be of much use, but if the game is fox or deer and a genuine working dog or bitch can be found the cross might be worth-while.

Deerhound or 'Stag'

In his own right, the deerhound is a crowd-stopper; to many sportsmen, especially those with any trace of the Celt in their blood, he is the epitome, the sum and substance of the dog; there is gracefulness, there is a presence, there is a wonderful flowing movement, there is a superb temperament; there is the hills, the heather, there is the hart, wounded by arrow or spear, and there is the dog to pull it down. All summed up in 100 lbs of grey beauty. Ah well . . . !

In the evolution of the running dog the larger, rougher-coated dogs were bred for hunting the deer, the wolf and the boar, the middle-sized dogs for the hare and the small dogs for the coney. The arrow and the spear are by no means instant killers and dogs were often needed to pull down a wounded deer. With the advent of guns, the range was short and accuracy was poor and dogs were still needed; the more so as communications became easier and others besides the Highlander discovered the delights of deer stalking. Many Highland lairds kept their breeds of dogs very jealously to themselves; they were never sold and seldom given as presents and, for many of the incomers from the South, the only source of dogs for stalking was from "doubtful henchmen of broken clans or poachers from over the hill with perhaps occasional help from passing Drovers". By 1825 the breed of Scotch deerhound had almost died out and many different crosses of lurcher were used for stalking. With the evolution of the high-velocity rifle the deerhound finally lost his vocation but modern breeders – to their eternal credit – have switched to coursing the blue and the brown hare with the very firm intention of keeping the deerhound as a working dog.

On the face of it, the deerhound might be the perfect lurcher: wonderful eyesight, nose enough to follow a blood-spoor, weather-proof coat, almost bottomless stamina, a wonderful temperament, and, if roused, a good guard dog. But he falls down in one important respect for work in the English countryside: his size means that he needs half a parish in which to turn and the rabbit is usually too small and nimble for him. Outside the deer forests he would not keep the pot filled, but I must admit a partiality – I like to see a touch of deerhound in a lurcher; say, an eighth deerhound in an otherwise greyhound-collie cross. The deerhound-collie cross can be very good if it is not too big but they often lack the dash that the greyhound gives.

In general, I am not very keen on the greyhound-deerhound longdog that one so often sees nowadays. Some are good, many are fast, but they don't seem to turn with a hare and I have seen a lot of them with a very odd action behind; a sort of hitching walk and trot as if they have something wrong with a hip.

"JESS"; a collie-greyhound cross

"FLYNN"; a collie-greyhound cross

Jim Meads

"WEDGE"; a deerhound-greyhound cross

"BONNIE"; an Old English Sheepdog-greyhound cross

Jim Meads

In the last century many deerhounds were exported to the Colonies; some families emigrating to Canada, America and Australia from the Highland clearances took their dogs with them. Deerhounds were sent to Australia for kangaroo coursing when the greyhound was found to be too small and they "turned out much superior to the heavy, mastiff-bred lurchers or light greyhound-like coursers that hitherto flourished along the Darling and Murray Rivers". Both pure and cross-bred deerhounds were used until recently in America for hunting coyotes and they are still used in Australia for hunting kangaroo and fox.

Foxhound
It is an interesting thought that if Harold had won the battle of Hastings and no other Norman invasion had been successful, there would, most probably, be no hunting in the British Isles as we know it today. The Saxons and their kin did not use packs of hounds for hunting by scent; the hounds, the customs and the language of the chase came from France.

This is not the place to discuss the evolution of the foxhound – though, in passing there is, apparently, evidence, or inference, that the Talbot received a cross of greyhound in the middle of the 17th century as a faster, galloping type of hound appeared in the North of England, whilst the slow Southern Hound was still used for some time in the South. The modern foxhound is the result of many generations of careful and thoughtful breeding, some lines tracing back for at least 250 years and probably earlier. It was not very long ago, before the necessity of using hound vans, that hounds were expected to trot 15 miles to the meet, hunt all day covering perhaps 40 or 50 miles, and trot back to kennels from where they left off, possibly a great deal further than their journey to the meet. They repeated this performance twice a week throughout the season, sometimes closer to home, sometimes further, and they carried on from November to April; and many hounds ran up with the pack for seven or eight seasons before age began to tell. A nose that can follow the faintest of scents down a tarmac road, stinking of diesel and rubber, courage to face the thickest of gorse covert and thorn hedge, activity to gallop and jump all day and stamina enough to wear down not only the fox but also the very best of blood horses; what more can a poacher want? There are, of course, drawbacks; the foxhound hunts by scent and not by sight; he is not fast enough to course a hare; the noise he makes is better than Bach to hunting people but would act like an alarm bell to a 'keeper; and he is a pack animal. All dogs were originally pack animals and the pack instinct is still present in most breeds as many owners of more than two or three dogs may well know to their cost. But the hound has lived and worked in a pack throughout the centuries and he cannot be expected to react and behave as other breeds do. What he does offer the lurcher breeder is strength, courage, stamina and nose but, as I have said about the Alsatian, when picking a puppy from a greyhound × foxhound litter, the ones that look like foxhounds when small will grow up cloddy and good for nothing.

It is interesting to read what those who have used the foxhound cross in the past have to say about it. William Scrope, writing in 1838, when he apparently had the freedom of the Atholl deer forest to kill whatever venison was needed for the Castle, to organise the drives for the Castle guests and to thin out the stock of deer when His Grace had "gone South" wrote: "Not being in possession of any of the celebrated race of the original Scotch greyhound which are now, indeed, very rare, and finding that all the dogs in the forest of Atholl were miserably degenerate, I bred some litters from a foxhound and a greyhound, the foxhound being the father. This cross answered perfectly: indeed I was previously advised that it would do so by Mr John Crerar [the head stalker] who after having tried various crosses for sixty years found this incomparably the best.

"Neither of these animals themselves would have answered; for the greyhound cannot stand the weather, and wants courage to that degree that most of them will turn from a fox when they come up to him and see his grin, and feel his sharp teeth; nay, they will scarcely go through a hedge in pursuit of a hare until after some practice. Besides, they have no nose and run entirely by sight; so that when the hart dashes into a deep moss or ravine, the chase is over and the dog stops and stares about him like the born idiot he is. The foxhound is equally objectionable; he has not sufficient speed, gives tongue and hunts too much by scent: in this way he spreads alarm through the forest; and if you turn him loose, he will amuse himself all day and you will probably see him no more till he comes home at night to his kennel. All these objections are obviated by the cross between the two. You get the speed of the greyhound with just enough of the nose of the foxhound to suit your purpose. Courage you have in perfection for most dogs so bred will face anything; neither craggy precipices nor rapid streams will check their course; they run mute and when they are put upon the scent of the hart

Jim Meads

"BUSTER"; a foxhound-greyhound cross

"JASPER; a staghound-greyhound cross

Jim Meads

Jim Meads

"RUPERT"; a greyhound–yellow Labrador cross

"GUSSIE"; a Labrador-greyhound cross

Jim Meads

they will follow it until they come up to him; and, again, when he is out of view they will carry on the scent, recover him, and bear the best greyhound to fits; I mean, of course, on forest ground.

"The present Marquis of Breadalbane had two dogs of this description, bred by me. They are now nine or ten years old (anno 1838) and his Lordship informs me they are still able to bring the stoutest hart in his forest to bay and are altogether perfect. These dogs in point of shape resemble the greyhound; but they are larger in the bone and shorter in the leg; some of them when in slow action carry their tails over their backs like the pure foxhound. Their dash in making a cast is most beautiful; and they stand all sorts of rough weather. As the above is, I think, the best cross that can possibly be obtained for the modern method of deer stalking, so it should be strictly adhered to; I mean that, when you wish to add to your kennel you must take the cross in its originality and not continue to breed from the produce first obtained. For, if you do this, you will soon see such alarming monsters staring around you as the warlike Daunia never flourished in her woods and thickets. . . ."

Scrope follows this with instructions as to how the dogs should be handled on the hill. This is not really part of my brief but it is well worth reading for those who have access to his book (1897 edn, p. 235 *et seq*). Scrope is, however, worth quoting again on the subject of giving the dog " . . . portions of the deer's liver when he is gralloched; but after having blooded them once or twice to enter them I do not think the custom should be continued, a dog's love for sport being independent of eating. . . . My objection to the system lies principally in the two following reasons: the first is that a dog can never run a second chase properly after having been so fed; the second, that when he has a deer in a wounded or dying state he is apt to help himself from the haunches before you have time to come up. A lurcher once damaged my sport in this villainous manner. I had wounded a deer which came out unexpectedly from Glen —— against my wind in a heavy mist. A dog belonging to the Duke was slipped and laid on the scent. For a long time we could neither hear nor discover the bay; at length we came suddenly upon it if bay it might be called. The dog had taken steaks from the living haunches, after the fashion of Abyssinians, and was amazingly turgid. His name was Hannibal. 'Expende, Annibalem, quot libras in duce summo invenies.' I gave him a pretty considerable drubbing but even under the lash it was some time ere 'La bocca sollevo dal fiero pasto quel peccator'. After this perpetration I changed his name from Hannibal to Cannibal; but I never suffered him to pass the Scottish Alps with me a second time".

It is interesting that Scrope mentions the resemblance of this cross to the greyhound. Others had different views on the greyhound-foxhound cross and Cupples quotes a Mr Robertson, of Black Mount Forest, as saying that: "Scropes' dogs were good but the deerhound-foxhound cross was better, being larger and rougher with more speed; but better still was the second cross deerhound-foxhound which was almost indistinguishable from the pure deerhound."

Modern life being what it is, one has neither the time, space or resources to try out the various breeding programmes that appear attractive on paper. There are several hound crosses that I would have liked to have experimented with, the fell hound, and the Dumfries hound being amongst them; when I was last at the great October Fair at Ballynasloe, some 25 years ago, I remember seeing what was said to be a greyhound-Kerry beagle cross; the greyhound blood was difficult to spot but perhaps I was "not told the whole of it". What a wonderful sight they were, those brightly painted tinkers' vans, drawn up round the sides of three or four fields.

Gun Dogs
Various gun dogs have been used from time to time as lurcher crosses; at one time a Flatcoat or Curly coat retriever cross was popular in the Midlands because of their size and 'go'. Labradors are too slow to offer much besides nose. Time and space are needed for hunting by scent; if the lurcher-owner has not got permission to be where he is then the sooner he concludes his business and gets off the ground the better. If he has got permission then a couple of bassets or beagles will look after the scent and longdogs can be used when the quarry is worked up to. If the only requirement is to work to a net then a retriever-collie cross or perhaps collie-pointer might be adequate, but, strictly speaking, these are not lurchers nor longdogs and, good though some of them may be, they do not concern us here.

Some time ago I saw a greyhound-springer spaniel cross that was said to have won its owner vast sums of money in matches of various sorts. However, when he came to let it loose after a hare at a coursing meeting the dog never got within turning distance of the hare and the owner was

Jim Meads

"SPINDLE' a greyhound–sloughi cross

"LUCY"; second generation from a Staffordshire Bull Terrier

Jim Meads

Jim Meads

"LINNET"; not very big but a
wonderful bitch after a hare

TWO GOOD SMALL ONES

"BRIDGET"; the Small Lurcher
Champion at Lambourn, 1976, and,
for the third year running, winner of
the small lurcher racing

Jim Meads

fined £5 for 'loose dog'. Perhaps Cotswold hares are tougher than those of the Midlands.

Saluki
A dog of saluki type has been in existence in the Middle East for at least 4,000 years and probably earlier. Isolated in the desert and carefully protected, treated as part of the nomad's family and sharing their hardships and occasional feasts the saluki has remained a pure breed; rigorous culling by the law of survival of the fittest ensured that no weaklings survived and, except for variations of size and feathering, there is probably little difference between today's saluki and that of 1,000 years ago. The saluki was introduced into England about the turn of the century and the Saluki Coursing Club, a branch of the Saluki Breed Club, has been running coursing meetings for some 40 years; their object is to ensure that the saluki remains a working dog as well as one that appears in the show ring.

The saluki is not very fast out of slips nor has it the speed of the greyhound; but it does have tremendous stamina and will gallop on at 'fourteen annas' until its prey is exhausted. The unkind might say that this is because it never goes flat out. The snag from a lurcher breeder's point of view is that a saluki can almost never be trusted completely with farm stock. I do not like the cross myself but that is a personal opinion.

Whippet
There are many theories about the origin of the whippet, most of which are variations on the theme of small greyhound crossed with terrier and Italian greyhound. To each his theory; but a superficial look at some of the major art galleries will show that there have been dogs of whippet type in Europe for at least 400 years. I do not believe that the whippet is the product of crossing anything and I do not believe that the whippet is a recent production. He has been here for a very long time and much in his present shape. There is a 300-year-old mention of "the little whippet or house dog". Outcrosses have certainly been made, just as they have with the greyhound; Lord Orford's breeding experiments introduced both Italian greyhounds and bulldogs into the ancestry of the modern greyhound and from the way that many whippets kill rabbits and hares – not to mention rats – there probably is a touch of terrier not too far back. In the days of rabbit coursing – two dogs slipped after a

rabbit that was loosed down the field in front of them – great strength was needed as well as speed since the match might be for the best of more than 20 rabbits. An outcross to bull terrier – not the roman-nosed white dog of today's show ring but to the fighting dogs of the Midlands – added this strength and courage, and some strains of whippet still have a look of the Stafford bull terrier about them.

Disregarding the purely show-ring whippets, many of whom could not, in Facey Romford's words, "catch a cat in a kitchen", there are a lot of very good, pure-bred coursing and working whippets that will do everything that a small lurcher will do and in many cases will do it better. Whippets cannot be seriously considered as hare *catchers* (as distinct from hare coursers); I have watched too many courses in public, both as owner and meeting secretary, to have any illusions on this score. But whippets are wonderful dogs for all forms of rabbiting, particularly when combined with a terrier or two; hedgerows, ferreting, netting, driving, the whippet will do them all. For this reason I do not see any requirement for breeding small lurchers since a working whippet will do all that the small lurcher will do and do it better; the only advantage that a whippet-Bedlington or whippet-collie cross has over the pure-bred dog is a weather-proof coat. Such a lurcher is almost always too slow to come up to a hare unless he is slipped very close indeed.

The only problem for the person wanting a working whippet is to find a working kennel from which to buy, but this is not difficult. The Whippet Coursing Club was started in 1963 and sufficient generations have been running in public to show which lines have speed, which lines have speed and stamina and which lines have both speed and stamina and, more important still, the hunting instinct. The breeders to go to are those whose dogs combine the very strong hunting instinct – by no means present in all whippets – with speed and stamina.

CHAPTER SIX

The Lurcher at Work – Rabbits

"No hunter of the age of fable
Had need to buckle on his belt,
More game than he was ever able
To take ran wild upon the veldt,
Each night with roast he stocked his table,
Then procreated on the pelt.
And that is how, of course, there came
At last to be more men than game."

In discussing the lurcher at work I shall describe some of the animals both in the British Isles and abroad that have been hunted or are now hunted with lurchers or longdogs. I take no sides on the rights or wrongs of such hunting; nor do I pass moral judgement on the uses to which lurchers and longdogs are put. So far as I am concerned, what a man does with his dogs is a matter for his individual conscience; all I do say is that, whatever the quarry to be hunted, it must be killed quickly and cleanly. If this cannot be done then the person concerned has no business to be out with lethal dogs. And, let there be no doubt about it, good lurchers are lethal; their purpose in life is to kill on their own, or to hold a large animal so that it can be knifed or shot.

In this, and the following Chapters I have listed animals in ascending order of size, from rabbits to red deer; sambur are included after red deer, as the stories about hunting them are quoted from Sir Samuel Baker, who I also quote on coursing red deer. Overseas, I have used longdogs for hunting jackal and blackbuck, and several correspondents describe the various dogs used in Australia and the quarry hunted with them.

A lot of lurcher owners will laugh at any mention of using quick-release slips; their dogs either run loose or are slipped by loosing a rope or lead or handkerchief from the collar. To each, his own method; because I have done a great deal of orthodox coursing I prefer to use single or double slips. Unless it is very highly trained, the dog that is running or walking loose beside its owner will start off after anything that moves, no matter how far away, and by the time a suitable hare gets up the dog is probably exhausted. Unless the object is merely a country ramble all dogs are better on leads. If the dog is loosed by letting go a rope or lead through the collar it means that the dog is running in a collar and, to me, this is the equivalent of working a terrier un-derground with a collar on; in the thickly hedged and wooded country in which I hunt, a dog sooner or later gets hung up if run in a leather collar. There is one exception to this and that is when one is deliberately out after foxes; the collar can save some damage to a dog's throat.

If there is any question of deciding which is the better of two dogs in a course then the only way of ensuring that they get away together at the same moment and on equal terms is to use quick-release slips; and, as I have found so many people who have spent years working with longdogs who have never used, or even seen slips, I make no apology for giving a description of how they work.

There are various different models of slips but the basic action is the same in all of them: the collars are held by spring-loaded pins which cannot be released until the slipper lets go of the lead and takes the weight of the dogs on the release cord. In the examples on page 59, a leather lead in the form of a tube (A) is fastened by a swivel to a metal tube (B), through the middle of which is a metal pin, held down by a spring. The collars (C) are fastened at one end to a 'D' (D) and at the other end of each is a metal tag (E) with a hole drilled through it; the pin inside the metal tube (B) fits through this hole and holds the collar closed. A cord (F) passes down the middle of the leather lead; at the top end of the lead is a wooden handle or dowel (G). The cord passes through this handle and is sewn into a leather wrist-strap (H) and at the bottom end of the lead the cord is fastened to a ring (J), at the top end of the release pin. To use the slips, the dogs are fastened side by side in the two collars; the slipper puts the leather loop round his wrist and holds the lead by the wooden handle. On seeing a hare the two dogs strain forward putting their weight on the collars; the slipper lets go of the wooden handle, thus taking the weight of the dogs on the leather loop round his wrist and so, on the cord down the middle of the lead. The weight of the dogs lifts the spring, the pin is pulled out of the slots

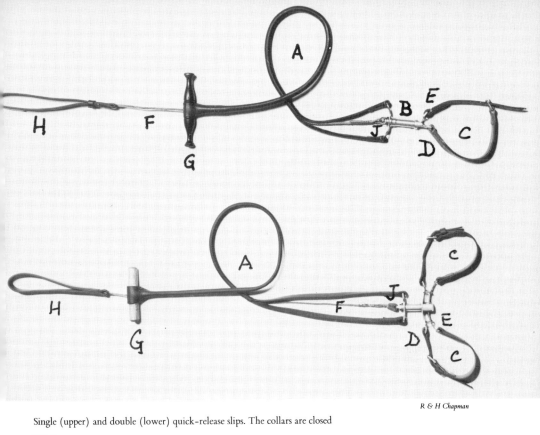

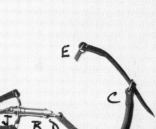

R & H Chapman

Single (upper) and double (lower) quick-release slips. The collars are closed

SLIPS

Single and double slips with the collars open after the dogs have been released

R & H Chapman

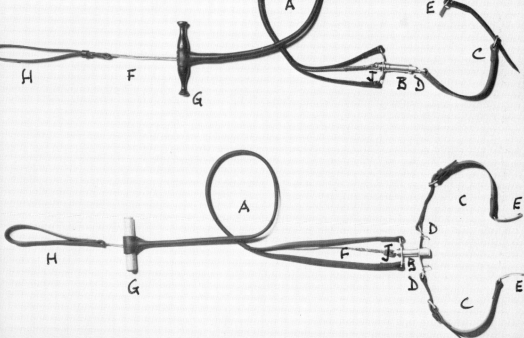

in the metal tags, the collars fly open and the two dogs go away together.

Embarrassing experience has shown me that, when out with lurchers, a knife is never so essential as when it has been left behind. I therefore always carry one. There are many different sorts of knife, but the blade should be at least three inches long and it *must* be sharp; Puma make very good knives.

And so to the first and smallest of our prey.

The Rabbit

The rabbit is not quite such an original inhabitant of the British Isles as the badger, but his bones have apparently been found in deposits dating from many thousands of years BC. He was not common until about 700 years ago and first mentions of him in documents appear about AD 1300. To many of us the wrong side of 50, he is the ubiquitous, prolific, rapacious, dodging brown streak that we cut our shooting teeth on. Provider of hilarious and exhausting sport when the last strip of corn was cut with the reaper and binder – but once described by a distinguished personage as being "six inches too short", and, at the speed with which he crosses a ride, it is easy to shoot just that distance behind him. He provided a steady income for the rabbit catcher, who was careful to leave a nucleus behind against his next visit. And he provided protein, thin leather, glue, fur and felt for various trades. He is almost impossible to eradicate completely by human, canine or mechanical means. Terrible stories are told of rabbits that swarmed on farms, whole fields that appeared to move, thousands of tons of produce ruined. But the truth was that the farmer who didn't intend to be overrun by rabbits wasn't overrun; only the man on whose boundary was a warren over which he had no control had a genuine grievance.

It took the virus of myxomatosis, that revolting disease deliberately spread by man, to bring the rabbit near the point of extinction in England. One only hopes that those responsible for its introduction and spread will spend the rest of eternity shivering in a sub-zero Hell – much worse than roasting. But even myxomatosis did not completely do away with rabbits. They survived here and there in twos and threes and gradually began to breed again, living above ground almost as if they knew that the rabbit flea was the vector of the disease and was waiting in the old buries. In my area the disease comes back in roughly a two-year cycle, but it seems that more and more rabbits are surviving each attack, many with the bare patches round the eyes of the rabbit that is now inoculated. They are back in reasonable numbers and a lot of fun can again be had with ferret, net, and dog. Ferreting does not come within the scope of this book and anyway there are three very good pamphlets on the subject for those who wish to start from scratch. They are the British Field Sports Society's *Rabbiting and Ferreting*, the Forestry Commission's Booklet No. 14, *Rabbit Control in Woodlands*, and the Game Conservancy's Booklet No. 22, *Rabbit Control*. Small lurchers can be used with ferrets but the only advantage that they have over the pure-bred whippet is a rougher coat to withstand bad weather.

For me, the best rabbiting for the lurcher is working hedgerows with a mixed pack. The composition of the pack is a matter for individual choice, but two lurchers, two whippets and two terriers should be manageable. Much as I admire the Lakeland and Border terriers when I see them working with Fell Packs, for hedgerow work I prefer terriers that are mainly white so that I can spot where they are in the hedge. The whippets will work both in the hedge and out of it, and the lurchers soon learn to keep level with, or just ahead of, the action; they also learn to nudge a bolting rabbit out into the field, away from the hedge, to make certain of picking it up. Main buries of course have to be ferreted, but there are always the outliers, and hedges should be taken away from the main bury so that rabbits bolt back. It is surprising how far they will go across a field rather than enter a nearer hole; perhaps they have their own 'Keep Out' signs.

Many people do not know the trick of getting a rabbit out of a shortish 'stop' without digging: cut the longest bramble you can find in the hedge, trim about a foot of thorns from one end for a hand-hold, push the bramble into the stop until the end is reached and then twist. If the rabbit doesn't bolt immediately it can usually be pulled out.

It is worthwhile carrying a couple of purse nets in one's pocket when 'hedging'. There is often an occasion when a purse net, tactically placed, will catch the rabbit that is bolting down the middle of a hedge; a known run across a gateway can be blocked, or a drain netted as one passes it.

If there is time and opportunity, stinking out is worthwhile. A bucket of sawdust or peat is mixed with creosote, diesel, Reynardine or any other really strong-smelling substance; enough to damp but not soak the sawdust. With a spoon tied to the end of a stick, a spoonful of the mixture is put into each visible hole; it is usual but not essential to block up the holes afterwards. Go back two days later and the rabbits will be sitting out, and good sport can be had.

I've used long nets but I've never had a lurcher well enough trained for the job and, unless a dog is very good, it will merely add to one's problems. There are enough things to go wrong when long-netting without adding to them unnecessarily. As good a method of moving rabbits is for the third member of the party to walk the field, slowly and quietly, rattling a box of matches occasionally.

Lamping

Coursing rabbits – and sometimes, hares – at night with the aid of a lamp can be very productive if conditions are right. The last time I did any lamping was before the war. My 'lamp' was the headlights of an Armstrong Siddeley, four-cylinder, open tourer. It was by no means a new car when I bought it for £25 in the mid-1930s, but on a straight road it could keep up its top speed of 45 miles per hour and, with a following wind and a long downhill slope, the speedometer had been known to approach the 50 mark; stirring stuff! 'Stirring' also applied to the gearbox; no synchromesh nonsense there. It was a point of honour with my brother and me that the clutch was only used for moving off from standstill, and gear changes had to be timed by ear; double-declutching became an art and gear-crashing meant a fine of a pint of beer. It was a strong car and on the South Wiltshire Downs it provided a good gun-platform for a shot gun, and could hold four or five longdogs, which were slipped 'over the side'.

The technique was simple; drive slowly across a field (one of those enormous downland pastures between Mere and Warminster, cropped close by sheep), switch on headlights: no rabbit in view, switch off headlights and, "driver, advance". Carry on for 200 yards, switch on headlights: rabbit in view, bang on the outside of the door to get him moving and slip a dog – or dogs. Then came the fun. The car of 40 years ago was a tortoise across country compared to a jeep or Land Rover and keeping up with the chase was as much a matter of luck as good judgement, particularly in a field full of grass-covered ant hills. But the downland rabbit was usually making for the edge of the down, or for a valley, where enormous buries were dug into the chalk hillside, so one could anticipate the probable run; the car had no roof so one was saved from concussion, the windscreen lay flat on the bonnet and fences were a long way apart. If a dog was approaching barbed wire or sheep netting

Jim Meads

The equipment: a rucksack frame is adapted to carry a lightweight battery and a "rabbit sack". The spot lamp has a thumb switch and the cable can be disconnected at the two-pin socket

LAMPING

The Operator with a dog "at the ready". Despite several attempts it was found impossible to take satisfactory pictures by the lamp beam at night so this photo had to be posed in daylight

Jim Meads

without a chance of catching the rabbit – or hare – one had to switch off the headlights and hope that the dog's night sight did not adjust instantly. On those great open spaces we did not have many dogs cut about at night; the only danger was if the rabbit turned back towards the car. Then one had the alternative of stopping and losing whatever was in the light beam or carrying on and avoiding the dogs. A spot lamp would of course have been the answer. I don't remember any very large catches but we had a lot of fun.

For various reasons I have not done any lamping since the war. With the advent of lightweight batteries and powerful spot lamps, I am completely out of date and I am most grateful to George Smith for the following account of lamping today.

"It was written by the great Egerton Warburton many years ago that 'Hunting's a science and riding's an art'. This applies to all sports including lamping (the pursuit of game with longdog and spotlight). Light, wind and weather conditions generally, are of paramount importance; the right dogs, people and equipment are no less important. Lamping in moonlight is a complete waste of time, as rabbits are always near home and seldom venture into the open far enough to make it anywhere near successful. The blacker and windier the night the better, with the exception of hoar frost (which is associated with moonlight), as neither rabbits nor hares like getting their feet too wet and cold. An exception to this rule is after a spell of prolonged rain on a dark windy night when rabbits and hares seem to be everywhere. Lamping, in the main, is confined to rabbits, but with sufficient science and the right dogs, hares and foxes will provide exceptionally good sport.

"There are exceptions to every rule, but greyhound-collie, small lurcher types, are the ideal dogs for lamping. Modern methods have enlightened us a lot, for my earliest recollections of lamping are pre-war days when we carried a 12-volt battery (borrowed from one of the few local car-owners) on our backs. The weight, as I remember well, was somewhere about half a hundredweight! These days, with modern lightweight batteries – English ones always preferred, of course, to Japanese – it is a different story. Quartz iodine and quartz halogen lamps can be very efficient, but they do tend to milk the battery. There are plenty of other sealed-beam unit types equally effective. One does not have to be a skilled electrician, but the advice from one does help in fixing up a set of tack. It is also a simple operation to make up a one-amp trickle charger, which prolongs the life of the battery, as there

Peter Goulding

A Norfolk rabbiting man

is always the danger of overcharging batteries, particularly this light variety. Everyone has their own idea of what a lurcher is, so does everyone have their own ideas as to what a good set of lamping equipment is. What does it matter so long as it suits the individual – sport is what we're after.

"Most lurchers take to lamping as they do to coursing or rabbiting in the daytime – practice is the only necessity in this respect and patience and perseverance on the part of the owner. Let them have a few 'easy' ones first and with practice they will soon become as crafty as the rest of them – and in a surprisingly short time as well. There is an element who believe that lamping is detrimental to dogs that are also used in the daylight; this is rarely so, and a good all-rounder is a good all-rounder, night or day; although undoubtedly, lurchers being what they are with regard to cunning, some may shirk a bit in the daytime. The man who coined the phrase 'Dogs are like people' must have had lurchers in mind when he did so, and I think that lurchers with the right amount of work and opportunity can become as cunning, or perhaps more so, than any of the other breeds.

"No-one can arrange a night's lamping with a friend 'on Monday at 8' or 'Friday at 7'; lamping must be done 'on the night' due to varying weather and light conditions. Soon after dark on a good, black, windy night perhaps, or after midnight on another night – one can never tell; but the tendency is always in favour of the dark, moonless night. Wind always helps, and remember to travel upwind to the quarry. Hospitality is a great thing, but don't be too hospitable in taking too many guests on lamping excursions. Smoking, or any form of noise such as coughing, talking, clanging of opening and shutting gates, are as detrimental as the moonlight when brer rabbit sees us from afar. With too big a team he hears us too soon and the result is the same – by the time we have got there he has vanished.

"A good dog will either catch and retrieve his rabbit, or will lose it at the covert or bury or wherever it escapes and will return immediately to his handler. A dog that 'hunts on' can spoil a lot of sport in frightening off any other rabbits that might be out and have not yet been picked up with the lamp. With comparatively little training the dog will run up and down the light beam like a yo-yo on a string, but, here again, perseverance is essential in getting him to do so in the first place. Always return your lurcher to slips after a course, before starting the next. A team of two participants, or three at the outside, is permissable

– two is ideal and three is approaching a crowd. Two men and two dogs make a good team as the dogs can be slipped alternately, ensuring a fresh one on each course and this does prolong the evening's sport. The rabbit will always run for home, the covert, bury or railway embankment, whatever affords him his most secure refuge. The vantage point of the operator, the one with the lamp, is of utmost importance, as he must illuminate the course from start to finish. It is best that the rabbit is running for home when the dog is slipped, as often a rabbit is facing from home when found. This eliminates any unnecessary turns. As in all things, practical experience can better explain this than any text or all the books that have been written.

"An important feature of lamping, especially in the early part of the season – September and October in particular – when long grass and stubble is plentiful – is the golden rule to go steady and take plenty of time; rabbits can go down and easily be passed over as, more often than not, the glint of the eye is the only indication that they are abroad. Rabbits can, when needs be, sit very tight indeed and you can be certain that when he does decide to run the gauntlet he goes straight into top gear; and in most rabbits, overdrive is also fitted. He is seldom far from home and he knows the way like the back of his paw. But, like the hare, he is a mixture of brilliance and stupidity and he can sometimes be persuaded into committing hara-kiri; should he enter a hedge or bush, the operator should keep the light on the spot without a move and the experienced lurcher will often push the rabbit out again.

"In my experience, farmland adjoining woodland is the most profitable area for lamping. Rabbits always run to covert and the operator walks about 30 yards out from, and parallel to, the covert with his assistant another 30 yards further out; in this way the most likely rabbit area can be covered with least effort and most chance of a large bag. Long slipping is a strictly daytime practice, and in lamping, the dog should not be slipped until it is certain that she is sighted and can get on terms with the rabbit. A few long and unsuccessful slips early on can take the edge off a dog and it is useless later on when one suddenly finds a lot of rabbits. One can never tell with lamping; after a blank evening the three fields right by the farm will be swarming with rabbits and the cry is 'I told you so, we should have come here first'.

"Lamping can be a most enjoyable sport; that is why most of us do it. But it can also be very lucrative and in undulating country a lamp is not nearly as conspicuous as one might imagine. In pre-myxomatosis days on a good night, one could easily take 120 rabbits with three good dogs; these bags have not been seen for some time but rabbits are back in numbers again in some parts of the country and cleanly killed rabbits are fetching good prices.

"Although far more arduous for the dogs than coursing rabbits, hares also provide very good sport at night, particularly in large fields; under such conditions they often get clear away from the lamp beam and the dogs return with clean mouths. But in small enclosures one often finds that hares, unlike rabbits, will not go through hedges or over walls that are lit up by the lamp beam, even to the extent of turning back into the dogs' mouths.

"As regards the breeding of lurchers for lamping this, like so much to do with lurchers, is a matter of opinion. But the best lamping dogs I have ever had were 'Lass', who carried Staffordshire bull terrier, whippet and greyhound blood; 'Spring', a Bedlington-greyhound; 'Fin', who looked deerhound-greyhound but had everything else besides, and stood nearly 30 inches at the shoulder; and one pure greyhound. These four would have beaten the world. As I have said, many people favour the smaller greyhound-collie for lamping but the truth is that the really good dog will do his work, no matter what he looks like."

There, indeed, speaks the voice of experience. Some may differ on a matter of detail, but it is what is carried back to the car at the end of the evening that proves the method.

"The darker the night and the more inclement the weather, the better the exercise"; no, not 'Monty' but General Ironside, at Aldershot in the 1920s. It certainly applies to lamping, and it is only choirs who know nothing of the game that sing "It's my delight on a shiny night" as the chorus to 'The Lincolnshire Poacher'!

CHAPTER SEVEN

Hares

As the Almighty created the Blackbuck and the Cheetah, the one to hunt the other, so He created the Hare and the Lurcher; and by this beneficent act He ensured, for some of mankind, an aeon of sport combined with a lot of succulent meals.

It is astonishing that the hare has survived to today. She is born in the open, she lives her life in the open; she has no natural refuge, she does not go to ground (though this can occasionally happen), she does not climb trees, she does not fly, she is a substantial meal for every predator in the countryside and yet she survives in large numbers. Her secret is that she has the 'prey' animal's reaction to danger, developed to a very high degree. Day and night she is ready at any moment, no matter what she is doing, to leap at one stride into full gallop. Her reaction is usually quick enough to avoid the feral cat or the stoat; on occasion, in defence of a leveret, she can deliver a kick that will send a cat head over heels. So long as she is fit, her speed is sufficient to outrun the fox. Her only real danger is the lurcher; not when coursing under rules, with the hare given a 100-yard start, a start that is often enough to take the edge off the pure greyhound's speed; but when the lurcher is being used to kill hares and is slipped at the moment that the hare gets up, no matter how close. Then her only defence is her ability to stop and turn from full gallop and to go on turning and turning, gradually nearing the gap in the hedge or the cover, be it a crop or a spinney, that she is making for. It is this constant turning that tires the dog, and, together with the long slip, is the reason why so few hares are killed at coursing meetings where the greyhound follows in the hare's track, and is penalised if he cuts the corner; 'nasty lurching brute'. The lurcher, on the other hand, who has not quite got the speed of the greyhound but from his cross breeding has, or ought to have, more brains, does cut the corner; when two dogs are used to working together one will often turn a hare into the other dog's mouth.

Taking the seasons round, most hares are killed by lurchers when 'out for a walk'. Two or three people walk a farm or several farms, dogs walk loose or are slipped by various means, hares are given no law and those that get up early escape whilst those that sit tight and get up at the last moment are caught. This method is particularly lethal just after harvest, before the stubbles are ploughed, or in the late winter when the corn is about six inches high, or the rape is starting to grow. Under these conditions good lurchers should kill seven out of ten hares and possibly more; it is a good method of hare control in that the really fit hares mostly escape to breed, the weak hares are killed and, unlike a hare shoot, there are no wounded hares to get away. They are either killed or go free and untouched.

Sometimes the farm walk is done by one person with his dogs, sometimes it is a friendly get-together of a dozen owners. A typical meeting might be eight or ten people with some 20 lurchers, some local, some from two or three hours away. They meet at a farm and move off when all have arrived. The line may cover 400 yards if the fields are big or may be as short as 50 yards if conditions dictate it. With a short line, probably only one pair of dogs will be slipped at a time from the centre of the line, if the line is a wide one the arrangement is likely to be that a pair are ready at either end of the line for the nearest hare that gets up. Inevitably there will be the occasion when a hare gets up in the middle of the line and either end think it is theirs; in fact four dogs are more likely to muddle each other than two dogs and that hare often escapes. The walk will go on until all dogs have had enough and then the adjournment is for lunch and to talk over the day's sport.

Those who have only seen the care taken of greyhounds at formal coursing meetings would be surprised at the casualness with which lurchers are treated. With greyhounds, two dogs are slipped, the course is run, the judge gives his decision and the handlers run to pick up their dogs; sometimes this is done quickly, sometimes one has the spectacle of the handler coaxing and entreating and throwing his cap along the ground to try and persuade a reluctant dog to come to hand. The urgency is all the more in that the dog must not get away after another hare. If the hare has been killed it may take all the longer to catch the greyhound that is trotting around carrying it. When the dog has eventually been caught his coat is put on, he is taken back to the dog van, his feet washed, his mouth washed out, he is examined for damage, rubbed down with embrocation, mouth washed out again and he is put

"They're running"; two greyhounds are slipped

COURSING UNDER RULES

'White collar" brings his hare round. Note the angle of the dogs' bodies and the strain on feet and tendons at 35 m.p.h.

Two whippets are slipped

Jim Meads

COURSING UNDER RULES

Whippets: the Judge, the hare and the dogs

Jim Meads

into the van on a straw bed to rest until wanted for the next round. With lurchers, if the hare is killed one of the dogs will probably bring it back, though a lot of lurchers will only retrieve when on their own or with a known companion. If the hare has escaped the dogs will eventually return to masters where their reception may vary from "Good boy, Darkie" to "C'm 'ere y' stupid b - - - - r"; a quick look to see if there is any damage, collar and lead are fastened, the line moves on for the next hare and as soon as those two dogs have their tongues in again they are ready for another course.

There are various forms of competition with lurchers, the commonest being "the best of three (or perhaps five) hares", each dog having to kill his hares single handed. A straight-forward form of betting that takes no mental effort and is open to every shade of one-up-manship, arguments and, occasionally, fights. The owner of the dog that is next to run makes certain that he sees the hare that gets up 35 yards away but when a strong hare starts at 100 yards he happens to to be looking the other way; if his dog has had a pounding and there is another hare to go, the dog, by chance, has a thorn in its foot which takes ten minutes to remove; and by that time the dog has its tongue in again. The permutations are endless and any such contest can only be unsatisfactory in that each dog will, by the nature of things, run different courses. It seems that the only really satisfactory way of testing one dog against another is to copy the Rules of the National Coursing Club (or Whippet, Deerhound or Saluki Coursing Clubs). It means that a minimum of formality has to be imposed but in my opinion this is better than the chaos that some lurcher days slide into: owners wandering off in all directions, loose dogs everywhere, and often three or more dogs chasing a hare.

Coursing under Rules is judged by awarding points for the work done by each dog during the course, the winner being the dog that scores most points. A course ends when the hare is killed or when the hare escapes and the dogs are unsighted. Under Rules the dog that kills the hare is not necessarily the winner; the other dog may have scored more points during the course. Points are awarded for:

the RUN-UP	for the dog that reaches the hare first, one, two or three points according to the superiority shown in speed;
a GO-BYE	when a dog, starting a clear length behind his opponent, passes him and reaches the hare a clear length

	in front, two points, or if gained on the outside of a curve, three points;
a TURN	when the hare is turned back from her line at more than ninety degrees, one point;
a WRENCH	when the hare is turned back from her line but for less than ninety degrees, half a point;
a KILL	not more than one point;
a TRIP	when a hare is thrown off her legs, or where a dog flecks, but cannot hold her, one point.

This is all straight-forward and the system has governed greyhound — and "other breed" — coursing for 150 years. But when we come to apply it to lurchers we find certain difficulties since the lurcher's purpose in life is to kill game and not waste time scoring points. Every course starts with the run-up to the hare, no matter how short; but a lurcher that is outstandingly fast and races on past the hare, instead of picking it up in his stride, is wasting his energies. Some lurchers will cruise up to a hare and follow it for a turn or two before making a proper run, so the dog that reaches the hare first may not be the better lurcher. Similarly with points awarded for turning the hare; the greyhound that turns and turns the hare is piling up points and even if he should stop with cramp or run out of steam his opponent may not be able to work off the points scored against him. But the lurcher that turns and turns the hare without catching it is wasting time and energy and should have points deducted from his score! Having caught his hare the good lurcher should retrieve it to hand so perhaps points should be given for a good retrieve; but many lurchers will only retrieve when on their own and if a strange dog is present will leave the hare where it was killed.

The truth is that the best lurcher is the one that kills the hare the quickest and brings it back to hand. But this pre-supposes a short slip, giving the hare as little chance as possible and is not sport but pot-hunting. For a day's sport the hare must be given reasonable law — at least 50 yards — and when this is done even very good lurchers may not kill as many as five out of ten hares, especially on heavy plough or arable.

This coursing will more resemble formal greyhound coursing and the rules of the National Coursing Club can reasonably be used to judge each course. It is not easy to get the judging right from one, static, viewpoint and at a formal meeting the only person who can really see what is happening is the judge on a horse who can stay close to where the action

Two deerhounds on fen-land

COURSING UNDER RULES

Saluki coursing in Oxfordshire

Jim Meads

The start

LURCHER COURSING

Lunch break

Jim Meads

is. It is very seldom that anyone is mounted at lurcher meetings unless, of course, the dogs are out with horses anyway; but if three people are nominated as judges; one at each end of the line and one in the middle, the chances are that the majority decisions will not be too far out in most cases. These three must have a working knowledge of the rules and there is a "trick" about adding up the score during a course. If the points awarded to each dog are kept separately it is very easy to forget the two totals in the general excitement. Imagine that a black and a fawn are to run next: the black is fast and gets up to the hare first, six or seven lengths ahead of the fawn (3 points to black), turns the hare back on her tracks (1 point to black, total 4), turns her on her tracks again (1 point to black, total 5), the fawn gets in and turns the hare (1 point to fawn, totals: black 5, fawn 1), then stays ahead of black for thirty yards and kills the hare (1 point to fawn, totals: black 5, fawn 2). An easier method of scoring this course is to say "black 3, black 4, black 5, black 4 (subtracting fawns 1 from black's total), black 3 (again subtracting fawn's 1 from black's total)". If the fawn dog had not killed but had gone on turning the hare without black getting in again the score would probably have gone "black 2, black 1, all square, fawn 1, fawn 2, fawn 3...". It is very much easier to remember one total than to try and remember two totals; using this method, all the judge has to do is to remember the total of the dog that has most points.

As regards the actual running of a lurcher meeting some hints may be useful to those who have not had to do it before. Firstly make certain — absolutely certain — that you have full permission, not only of the land-owner but of any tenant farmers there may be; hares are ground game and tenants have the right to them. If the land-owner or 'keeper is not coming with you make quite sure that you know the boundaries; in law there is a right to follow a hare across a boundary "in fresh pursuit" (Game Act, 1831, section 35) but in practise this can cause trouble. If hares have a known run off the estate or farm where you are coursing it may be worthwhile seeing the neighbours concerned before you start.

Before moving off from the meeting place everyone should be reminded not to slip on a hare that gets up under their feet but to give it reasonable law; and also reminded that it is usually a waste of time to slip on a hare that is far away, that is disappearing over the brow of a hill or that is near a hedge. The two dogs to be slipped should be led some 20 to 30 yards in front of the line; how they are slipped is up to their owners. Quick-release slips are the best but they are expensive now-a-days; many people use a string or strap through the collar. It is best if one of the handlers says "are you ready — slip" or something like that to try and get both dogs off together. It is worthwhile putting the next two dogs to run in the centre of the line so that they are ready when the time comes; if the line has to stand still for a long time waiting for dogs to come back a hare will sometimes get up and go in the opposite direction. Rather than "waste" a hare, the next two to run can often be slipped in these circumstances.

The line should not be too long: 100 yards is quite enough, 60 yards is often better and if there are are not many walkers the line should be even shorter, particularly in long grass or stubble or rough plough. If the gaps between walkers are too big many hares will sit tight and be missed. Try to take each field up wind or across the wind and stop people from talking; female voices will clear hares from the next field before you get to it. Finally I have a local rule that, once we start walking, there is fine of £1 for each loose dog, the proceeds going to the Coursing Supporters Club.

But when all is said and done, most of us do most of our coursing with lurchers just for the pleasure of seeing them run.

I wonder how many people still use - or even own - a gate net nowadays? With a dog that knows her business, and under the right circumstances, it can be a very productive method of catching hares; and although almost any dog of reasonable size could be taught to do the work, a lurcher, who has both the brains and the speed, is the proper answer. A field is selected where hares are known to feed and the gate is approached after dark. (The gate is opened if it is possible to do so quietly.) A net is stretched between the gate posts, to hang down with a good loose fold onto the ground in the gateway. The dog is sent out into the field and the owner crouches down in the ditch by the gate. From now on all depends on the lurcher. She trots quietly round the hedgerow to the far side of the field and proceeds to work the field backwards and forwards, like a pointer on a grouse moor, each turn taking her nearer the gate. If, and when, she puts up a hare she does not try to catch it but courses it at half speed towards the gate. If she sees that it is making for a smeuse in the hedge she puts on speed and nudges the hare towards the gate. If the hare is determined to leave the field through the hedge the lurcher accelerates and tries to pick her up and bring her back to master; but if all goes well the hare, seeing the gate is

Walking the stubble
LURCHER COURSING

An uneven slip

LURCHER COURSING

The two dogs to be slipped should walk about 30 yards in front of the line

open, puts on speed and smack! into the net she goes. The net, which is merely held up by a stone on each gate post, collapses, master is out of the ditch, the hare's neck is broken, the net is folded, net and hare go into the inside pocket of his coat and master and dog are on their way to the next suitable field.

Easy to write about, not so easy to do. The wind can be wrong, the field can be blank of hares, the lurcher can cut a corner, the hare can go out of the far hedge, the net can fall off the gate post. But on the occasion when it comes off it can be most exciting, particularly if it is in the dark. One would be crouching in the ditch by the gate, head down, so that one's face would not show up. All one has to go by is sound: a dog galloping? Feet swishing through the grass? That was a grunt as the lurcher turned the hare. The noise gets closer, left, then right. Now they're coming, and fast, too, from the sound of them. Gate? or that hole in the hedge that we forgot to block? Gate it is. Out we go as quick as possible; the dog wasn't able to stop and dog, hare and net are out in the lane. What a muddle to be sorted out in the dark.

It is not necessary for the gate to be open, but if it is shut the hare tends to slow down and may not go through fast enough to roll up in the net. One friend of my youth used a double net; normal mesh on the inside and a larger, eighteen inch mesh on the outer so that in theory the hare would go through the larger mesh and be held in the pocket so formed. The snag came when she had to be removed from the net in a hurry. On one occasion the village police sergeant (there were such people in those days) was not away for the evening after all and we had to run for it, one of us with the net, complete with hare, over his shoulder, in the manner of a Scots shepherd with a lamb in his plaid, but at a somewhat greater speed! Luckily the hare kept quiet, though by no means still, and we got away with it; very much more by luck than good management.

When it wants to, the hare can jump enormous distances. There have been instances of hares jumping the Aintree fences, and regular followers of hare hounds will have seen a hare take a standing jump eight or ten feet sideways and clap down. I have only twice seen a fox after a hare, once in a straight forward course, the hare hardly extended and the fox giving up after a couple of fields. On the other occasion the hare was surprised by a fox in a gateway. The hare jumped first, easily clearing the fox. One spring I was standing in the corner of a field watching three 'March hares' boxing and capering. One of the hares broke off the game

Slipping: the author's black puppy is well into her stride with the brindle just being released

LURCHER COURSING

Two dogs are slipped across the line on a hare that went off sideways

Father and son are slipped on a swinging hare

LURCHER COURSING

Watching a course

Running-up

LURCHER COURSING

A small lurcher doing all the work

Jim Meads

Turning on wet arable

LURCHER COURSING

"Puss" is away again

Jim Meads

A stern chase

LURCHER COURSING

Co-operation

The end of a successful course. Although the handlers are holding four dogs, only two dogs are ever slipped at a time. More than two dogs get in each other's way.

LURCHER COURSING

"Right: through the gate and we'll try the next field"

and cantered down the field towards me. When it was about ten yards away, two whippets, who had been busy in the previous field, came through the hedge, saw the hare and went for it head on. Without altering speed or direction the hare jumped over the two dogs, clearing them comfortably and kicking back as it did so. On a more recent occasion two whippets were coursing a large leveret between windrows of straw. The leveret turned and they brought it back up the next row to where I was standing with a lurcher on a lead. The leveret was watching the dogs chasing it and did not see us until the last moment. Instead of turning, it jumped and the lurcher caught it in mid-air.

CHAPTER EIGHT

Foxes

It is interesting to look at the changes in the social level of the fox in the British Isles, or, rather, in England. For a great many years he was treated as vermin, often with a price on his head and certainly not worth the notice of those who hunted the stag or the hare. There was a gradual changeover from 17th and 18th century hare hunting to foxhunting, followed by a period of adulation when vulpicide meant social death in some circles ... and now there is the Damocletian sword of rabies hanging over the fox's head; if rabies does ever break out in England there is certain to be a hysterical outcry for the fox's eradication. Little do those who call for such campaigns realise the problems involved and the unpredictable effect on the countryside. Other countries live with rabies and we may have to do so yet.

Hunting is said by its supporters to be the most efficient method of killing foxes. In certain parts of the country this is true, but as a general statement it is wishful thinking; stirring up the foxes and killing some of them, yes; but the most efficient fox killer is the 'keeper or farmer who does not intend to have foxes on his land. He is there winter and summer, he has the opportunity to shoot, trap, snare and gas; and he has no chivalrous thoughts about leaving a fox underground to "give a good run another time". On unkeepered land it is often the unofficial 'pack' that kills most foxes: men who know their ground, using dogs, nets and guns. I have had a good share of hunting; I was blooded with Lord Leconfield's in 1921, I was lucky enough to see the great 'Ikey' Bell with the South and West Wilts, I still watch hounds when I can, on foot, and I fully agree with the famous M.F.H. who said that most people go hunting because it is "bloody good fun"; but there are more efficient ways of killing foxes. If packs of hounds published their tally and it was possible to get an accurate total of the foxes killed by other means in each hunting country I believe that the figures would be illuminating.

Being omnivorous and highly adaptable the fox has benefited from the urbanisation and industrialisation of the country and has not only spread into surburbia and even town centres but has given sanctuaries in the shape of Forestry Commission woodlands, the vicinities of motorways and railways and other areas where hounds draw seldom or not at all. Fox control in these areas varies from the 'expert marksman' from some Ministry or other, through farmers' shoots, snaring and gassing, to the unofficial foxhunters with their terriers, 'smell dogs' and lurchers They account for a large number of foxes every year. There is no question of 'coursing' a fox; any lurcher that cannot give a fox half a field's start and roll him over before the hedge ought to be put down as he certainly won't catch anything else. Being a predator himself the fox has no automatic escape reaction to danger as the prey animals such as the hare and rabbit have. When put up, out of a hedge or bramble brake or from a pollard willow, he goes away slowly and almost reluctantly swinging his brush from side to side and looking as if he is uncertain whether he should have moved at all, let alone where he is going. It is different if the fox is driven from a patch of kale or a small spinney. Then he has had time to think and he usually goes off at best speed; but even so he is no match in speed for a moderate lurcher, and if the country is at all open a couple of lurchers should catch up with him very quickly. The end depends on how hard the lurchers are; the trouble is that a hard dog to fox does not usually hold back with rabbit or hare, and, even if such a dog will retrieve, the resultant carcase is sometimes only fit for dog food.

This is not a modern problem. Thomas Fairfax, in his *Complete Sportsman, or Country Gentleman's Recreation* printed in 1760, says: "In coursing a fox, no other art is required than standing close, and on a clean wind on the outside of some grove, where you are to expect his coming out, and then give him head enough, otherwise he will turn back to the covert: for the slowest greyhound will be swift enough to overtake him; all the hazard of this course is the spoiling your dog by the fox, which oftentimes happens; and for this reason you should not run any that are worth much at this chace; but such that are hard bitten dogs that will seize anything." Fairfax goes on to say: "If greyhounds course him on a plain, his last refuge is to piss on his tail, and flap it in their faces as they come near him; and sometimes squirting his thicker excrement upon them, to make them give over the course or pursuit." There have been times, when my dogs have killed a fox on a wet day and they are drying off in a warm room that, from the smell, I wonder if Thomas Fairfax was not a better naturalist than appears at first sight.

The fox that fights a couple of lurchers is sometimes not worth skinning and with fox pelts fetching up to £20 it is a rich man that can let that sort of tax-free bonus rot in a ditch. There is also the question of damage to the dogs, as Fairfax points out. A fox's bite, usually on the neck or lower jaw, or through the nose, can take some time to heal, even if washed out straight away with Dettol, and can put a dog off work for several weeks. I have found the best answer, if I get there in time, is to drop the dogs' leads onto the fox's head (or a purse net if I'm carrying one); he will usually snap at the leads and hang on, giving one the chance to keep his teeth off the dogs; a quick thrust behind the elbow with a knife ends the contest. The damage occurs when a fox is cornered or has a running fight with lurchers. The big, strong dog that breaks the fox's back in his gallop is, of course, not damaged.

A season or two ago I fell in with a 'gun pack' in central Wales. The four men had a van loaded with lurchers, terriers, beagles, shot-guns and a sack of purse nets. The first draw was a small field of kale; one side against an oak wood with little undergrowth, the other three sides onto open grazing and moorland. The two guns took the wood side of the kale and the other three sides were each covered by two lurchers – my own dog making the sixth – with the pairs of dogs held hard back against the kale. The beagles and two terriers were put in from the side opposite the wood and they soon opened. Not many minutes passed before a shout from the North side meant that a pair of lurchers had been slipped; almost immediately a fox came out on my side of the kale and about 40 yards from me. I gave him time to get into his stride and heading away from cover, and I slipped the lurchers; the strange dog I was holding was of a size and weight with my own bitch. They caught him in about 150 yards and by the time I got up to them there was no need for me to deal with him. Nothing else came out my side of the kale and soon there was a shout for us to pack up and go back to the cars; the bag so far, two foxes with skins hardly damaged. A ten-minute drive along narrow lanes brought us to a steep valley where, two days before, I had watched the local pack of Welsh hounds hunt a fox for an hour before he went to ground in a wood on the valley side. To my surprise the leading van parked just inside the wood and we walked up the hill to the same earth. Purse nets were placed over the holes, two terriers were slipped in past the nets like ferrets, the men with guns stood above the earth, those of us with lurchers were stationed at suitable distances up and down the slope, and we waited in silence. After a while one of the listeners at the earth held up a hand to indicate a terrier speaking; within minutes I saw the same man gesture at the two with guns, one of the purse nets erupted from the earth and a fox was rolling down the slope curled up in the net, closely followed by a second fox that jinked off across the slope. Events moved fast: the net was retrieved and the fox hit over the head, the jinking fox was shot, and a third fox came out into a net and was retrieved by two lurchers. My acquaintances thought that this was the lot, and by 12.30 we were in the nearest pub, with a total of three fox pelts, and two live foxes in sacks. I did not like to ask which hunting country they were destined for.

Sir Walter Scott has a good description of a fox hunt in *Guy Mannering*. Captain Brown is staying with Dandie Dinmont at Charlies Hope: "They had gradually ascended very high and found themselves on a mountain ridge overlooking a glen of great depth but extremely narrow. Here the sportsmen had collected, with an apparatus which would have shocked a member of the Pytchley Hunt: for the object being the removal of a noxious and destructive animal as well as the pleasures of the chase, poor Reynard was allowed much less fair play than when pursued in form through an open country. The strength of his habitation however and the nature of the ground by which it was surrounded on all sides supplied what was wanting in the courtesy of his pursuers. Along the edges of the ravine, the hunters on horse and on foot ranged themselves; almost every farmer had with him at least a brace of large and fierce greyhounds of the race of those deer dogs which were formerly used in that country but greatly lessened in size through being crossed with the common breed. The huntsman, a sort of provincial officer of the district, who receives a supply of meal and a reward for every fox he destroys, was already at the bottom of the dell, whose echoes thundered to the chiding of two or three brace of foxhounds. Terriers, including the whole generation of Pepper and Mustard ['Dandie Dinmonts'] were also in attendance, having been sent forward under the care of a shepherd. Mongrel, whelp, and cur of low degree filled up the burden of the chorus. The spectators on the brink of the ravine held their greyhounds in leash, in readiness to slip them at the fox, as soon as the activity of the party below should force him to abandon his cover. When the fox, persecuted from one stronghold to another, was at length obliged to abandon his valley and to break away for a more distant retreat, those who watched his motions from the top slipped their

Jim Meads

A hare at speed

"Charlie" on his way

Jim Meads

greyhounds, which, excelling the fox in swiftness and equalling him in ferocity and spirit, soon brought the plunderer to his life's end. In this way, without any attention to the normal rules and decorums of sport, but apparently much to the gratification both of bipeds and quadrupeds as if all due ritual had been followed, four foxes were killed on this active morning."

Here is Charles St John on the subject of foxhunting in Eastern Scotland, 150 years ago: "What little I have to say on this most momentous of all sporting subjects will, I fear, be sadly

'Unmusical to Melton ears
And harsh in sound to Quorne.'

But what are a set of poor fellows like us to do, living here amongst mountains and ravines and torrents and deep water courses against none of which the best horse that ever put foot on turf could contend for five minutes? It took me, I must confess, some time before I could get over the finer tone of my Leicestershire feeling and I have no doubt that I blushed a perfect scarlet the first time I doubled up a fox with a rifle ball. Of all the ways of earning a living perhaps there is none that requires a greater degree of hardihood and acuteness than the trade of a vermin killer in the Highlands – meaning by 'vermin' not magpies, crows and such 'small deer' but the stronger and wilder carnivorous natives of the mountains and forests – the enemies of the sheep and lambs. In the Highlands he is honoured with the title of the Foxhunter, or Tod-hunter; but the Highland foxhunter leads a very different life and heads a very different establishment from him of Leicestershire. When you first come upon him in some wild glen you are somewhat startled at his appearance and bearing. He is generally a wiry, active man, past middle age, slung round with pouches and belts for carrying the implements of his trade; he wears a huge cap of badger skin and carries an old fashioned, long barrelled fowling piece. At his feet follow two or three couple of strong, gaunt slowhounds, a brace of greyhounds, rough and with a good dash of the lurcher, and a miscellaneous tail of terriers of every degree."

And, to move forward 140 years, from 1836 to 1976, we are in time for a 'quick thing' with the A----y Fur and Feather Hounds. It is early morning in the autumn and hounds are out for some hedgerow hunting: "There's Spider marking along the hedge; she wags her tail for partridge or pheasant but she is still now so this must be a rabbit or a hare. Honey joins her and is into the hedge, out the other side where Ticket screams with excitement and is into the hedge 10 yards further on; the lurchers dart into the hedge and out again, down the field for 50 yards, skid to a stop and back again. Now the terriers are in the ditch; it's three foot deep and covered with brambles; this rabbit shouldn't last long. Honey gives tongue and out it comes – by God it's a fox! Tarn strikes but fails to hold him and he's into the ditch again; on the far side the whippets are shouting with rage 20 yards down the hedge, the lurchers marking this side. Out he comes, just beyond them and a bending race goes on in and out of the hedge. Every time a dog strikes he gives a flick of his brush and he's into the ditch. Silence: all the dogs this side of the hedge, which is a mistake. Terriers into the ditch again and 'Yonder he goes' – he's broken out the far side and is making for the Old Barn, 'Huic, huic, ha-la-la-la-la'. The lurchers are through the hedge with the whippets on their heels. The fox has a good start, 30 yards to go, 20 yards and Tarn rolls him over but fails to hold, the fox dodges away under the puppy, the two lurchers collide and the fox is into the far hedge.

"Now, is he in the big culvert or is he in the Barn and cattle buildings? The lurchers and whippets are in and out of the hedge where he disappeared so we must rely on the terriers: 'try here, try here' and they're into the culvert and through it. He must be in the buildings or we'd have seen him go away over the slope the far side. The terriers carry a line over the concrete into the Barn, through the cattle yard and into the straw bales. A whippet tries the old earth under the water trough but leaves it straight away. 'Lieu in dogs, lieu in' and they go backwards and forwards up and down the bales, hackles up and growling. A shout from Honey and a roar from Tarn and the puppy, a bale goes flying and they've got him; it's all over, a good-sized dog fox, looks like 15 lbs, nice pelt not badly torn, and very little damage to the dogs; a couple of torn faces and a bit of skin off a leg. Nothing that Dettol and green oil won't cure."

CHAPTER NINE

Deer

The lurcher and the longdog in their different forms, usually under the name of 'greyhound', have been used for hunting deer from time immemorial. Man has been on this earth in recognisable shape for some ten million years and only in the last 10,000 years has he changed from hunter to nomad, to agriculturalist, and finally to urban manufacturer. It is no wonder that the hunting instinct is still very much alive and that lurchers and longdogs are still used for hunting deer. That such practices may be illegal matters little to those who get away with it. Most of those who are caught, pay their fines and start again. For this reason I leave the details of coursing deer to three of the authors who have written the classical descriptions of it: Scrope, St John and Baker. But first it is worthwhile looking at the species of deer that are wild in the British Isles, and in doing so I repeat what I said at the beginning of Chapter Six: that what a man does with his dogs is a matter for his own conscience, but whatever killing is done must be done quickly and cleanly and those that cannot ensure this should not be out with lurchers.

The Muntjac, or barking deer, though not a native, is the smallest of the deer to be found in England. They are the descendants of wartime escapers from Woburn and have gradually spread outwards from Bedfordshire to neighbouring counties wherever there are woodlands. I believe they have reached the outskirts of London. Though quite common in places they are usually secretive and many of them are harboured unknowingly; in fact I am often surprised at the people who do not know what they are when they see them. A fully grown buck measures about 24 inches at the shoulder and weights up to 38 or 40 lbs; when cleaned and skinned they weigh about 23 or 24 lbs and make very good eating. They live in pairs, or singly, in low, brambly undergrowth through which they have their runs and tunnels. They come out into the open in the late evening or very early morning, but can occasionally be found moving from covert to covert at any time of day. They are usually killed by chance in woodland, but if met with in the open and some way from covert they can give a good course as they have a surprising turn of speed for half a mile or so. Two lurchers have little difficulty in catching and killing barking deer, but a mature buck has very sharp eye teeth similar to the tushes of a pig but pointing downwards and can cut up a single dog quite severely.

A barking deer can be quite a weight to carry any distance and as soon as it is killed it should be hung up to a fence or branch, and gralloched and the head cut off. The neck should be left as there is a lot of meat on it. When skinning a barking deer, it is interesting to see how much thicker the skin is on the neck, shoulders and chest than on the flanks and quarters; I imagine this is because they normally live and move about amongst thorns and brambles.

Roe and Fallow Deer
Next in ascending size are the Roe and Fallow deer; apart from the very rare visitor they are never seen in my area, but fallow are to be found at the other end of the county. Both deer are well distributed throughout Britain, and in places are plentiful. Roe are difficult to get near as any stalker knows, but Fallow deer are sometimes poached with dogs, both by day and by night, in the latter case the deer first being found by spotlamp; a wasteful and usually cruel method as the deer are often wounded or killed some way from where the dogs are first laid on and are not picked up. Even in daylight the course often takes place through woodland and nothing is seen of it.

Red Deer
With the red deer we come to the one British species which – living in open country – is, or was possible to course with the expectation of seeing something of the sport. I have never, to my sorrow, seen it done nor have I more than the slimmest chance of ever doing so; but two puppies from my last litter have gone to a very sporting home not far from the Great Glen and, who knows.... Somewhat naturally those who have coursed and do course red deer do not discuss such things; but there are many classical descriptions of the sport which make the modern lurcher-owner's mouth water. The purist may object that in several stories the coursing is done with deerhounds; however, the true deerhound was, apparently almost extinct by 1830 and, as these descriptions were written during the ensuing 30 years it is more than

The Muntjac, or Barking Deer; a buck in velvet. The tip of the canine tooth, that can cut a dog badly, can be seen projecting below the upper lip

Roe deer

John Marchinton

A Fallow Buck

The Noblest of All: Red Deer

G. Kenneth Whitehead

probable that, with the exception of the coursing on Jura, when Colonsay dogs were used (these being amongst the few pure-bred dogs left), the others were more or less cross-bred.

Here is a visit to Jura in 1835, by Archibald MacNeil of Colonsay.

"Deer-coursing, the noblest of all the Highland sports, has long been a favourite amusement with the inhabitants of the north and west of Scotland." The party, consisting of "six sportsmen, a boat's crew of seven men, with piper, deer-stalker, and two deerhounds" arrive on Jura on an August evening and make camp in a cave. One of the sailors acts as cook and "by the side of the fire was spread a couch of dried ferns and heather such as fair Ellen provided for King James; but although our attendant was neither young nor of the fair sex we had the advantage over royalty in one respect being provided with a good stock of blankets". The two deerhounds, Bran and Buskar, are fastened to a stone at the back of the cave "large enough to have secured even the hounds of Fingal" and the party sleeps till dawn when the piper wakes them with 'Hey Johnny Cope'. After a quick dip and breakfast, the party starts off up a rocky glen "for some miles" with the stalker about fifty yards in advance. Eventually he spies a stag about a mile off and everyone creeps back until they are hidden. "As yet the rest of the party had seen nothing of the stag and although the stalker pointed steadily in the direction in which he was, not one of them could discover him with the naked eye; but Buskar, who had hitherto followed quietly, now commenced a low whining, and, with ears erect, gazed steadily at the spot where the deer was lying." After a long stalk, mostly *ventre à terre*, the stalker brings them within 100 yards of the stag. "Buskar soon put the matter beyond question, for, raising his head, he bounded forward and almost escaped from the person who held him. No time was to be lost: the whole party immediately moved forward in silent and breathless expectation with the dogs in front, straining in the slips; and on reaching the top of the hillock we got a full view of the noble stag, who, having heard our footsteps, had sprung to his legs and was staring us full in the face at the distance of about sixty yards. The dogs were slipped: a general halloo burst from the whole party, and the stag, wheeling round, set off at full speed with Buskar and Bran straining after him. The brown figure of the deer with his noble antlers laid back, contrasting with the light colour of the dogs stretching along the dark heath, presented one of the most exciting scenes that it is possible to imagine."

The deer makes for some rising ground, but finding the dogs pressing him close he turns and retraces his steps, glissading down an almost sheer slope, the dogs following; "and the chase was continued in an oblique direction down the side of a most rugged and rocky brae, the deer apparently more fresh and nimble than ever, jumping through the rocks like a goat, and the dogs well up, though occasionally receiving the most fearful falls. From the high position in which we were placed, the chase was visible for nearly half a mile. When some rising ground intercepted our view we made with all speed for a higher point and, on reaching it, we could perceive that the dogs, having got upon smooth ground, had gained on the deer who was still going at speed, and were close up with him. Bran was then leading and in a few seconds was at his heels and siezed his hock with such violence of grasp as seemed in a measure to paralyze the limb, for the deer's speed was immediately checked. Buskar was not far behind and soon passing Bran he siezed the deer by the neck. Notwithstanding the weight of the two dogs that were hanging on to him, having the assistance of the slope of the ground, he continued dragging them along at a most extraordinary rate and succeeded more than once in kicking Bran off. But he became at length exhausted; the dogs succeeded in pulling him down and although he made an attempt to rise he did not regain his legs. On coming up we found him dead with his throat perforated". Some details are given of one of the dogs, Buskar. He was 28 inches at the shoulder and weighed 85 lbs. He was a pure deerhound and was pale yellow with wiry, elastic coat.

Charles St John has a good description of coursing red deer in his *Wild Sports and Natural History of the Highlands,* and is writing of about the same period as Scrope. St John is more of a naturalist than Scrope, who was an out-and-out sportsman, and he has his reservations: "Though I am by no means of opinion that running red deer with the rough deerhound is so exciting or so satisfactory a sport as stalking the noble animal and attacking him in his fastness with the aid only of a rifle, I have sometimes seen runs with the deerhounds which fully answered all my expectations. It much oftener happens however, that, after the first start, nothing more is seen of dogs or deer until they are found at bay in some rocky burn or stream, the whole run having taken place out of sight of the sportsman. Moreover, the dogs run a great risk of being disabled and injured either by the stag or by the sharp and rugged rocks over which they take their headlong course... the deerhound is in-variably a pet and favourite of his master so that any accident that happens to him is the more regretted. It will often happen that the dogs will set off after some hind or young stag who leads both them and you away a long chace, unsatisfactory both in its commencement and termination, disturbing the ground and taking up twice as much time as would be required to kill the fine old ten-antlered stag whose head you covet." St John quotes a price of 50 guineas as not being unusual for a first-rate dog while 20 to 30 are frequently given for a tolerable one. Translated into today's equivalent, 140 years later, these are sums that many breeders would like to lay their hands on.

Whilst Scrope tells us a great deal about using dogs to follow a wounded stag, for sheer professionalism and experience of coursing deer we must go to Sir Samuel Baker, that great traveller, hunter and observer of wild animals. Writing in the 1880s he describes a trial at Blair Atholl: "There is hardly a more sporting sight than a stag at bay; but as the dogs are trained simply to follow a wounded deer until it stands, the termination of the hunt is a tame affair as the deer is shot directly that the rifle arrives on the scene. About thirty two years have passed away since we discussed the question whether the deerhounds at Blair [which Scrope tells us were cross-bred] would seize a stag if it were necessary. Most people who knew the training of the dogs thought not. The Duke of Atholl inclined to that opinion. On the other side I thought they would, provided that no rifles were taken out, and the dogs should see that the stag was to be tackled at close quarters with the knife. There was never a keener sportsman than his Grace and he was good enough to consent to a trial. It was arranged that I might select any two of the deerhounds and hunt down a fresh stag, bring it to bay and kill it with a knife. To myself the affair appeared exceedingly simple as I had been accustomed to this kind of hunting for many years in the mountains of Ceylon, but the others disbelieved that the two hounds would bring a fresh deer to bay as they had always been accustomed to follow animals that were wounded. We were a large party and we met at ——— about 10 miles from the Castle."

There follows a description of the Glen and the autumn tints. "The afternoon was perfect; all that was required was game. The presence of many ladies brought us luck for, after scanning in vain a long expanse of country we were suddenly delighted by the almost magical appearance of a stag; he had been lying down behind a large rock a little more than half way up the hillside. He was about 1,000 yards distant and he stood regarding the carriages and our party, which included the keepers and the two hounds. Fortunately we were on the main road upon which the deer were accustomed to see travellers (although few) who did not interfere with their domain. It was arranged that the party should drive back a mile towards the Castle, while I should walk on and discover a favourite position for ascending the hill and coming down from above upon the stag. The party turned back and I continued on my way accompanied by two of the hill-men and the dogs."

They climb the hill and on coming out above where the deer had been standing there was no sign of him. However the gillies think that the stag lay down again as soon as the carriages disappeared. "In this opinion I agreed; we accordingly held the dogs in readiness to slip and the gillie led the way. There was no doubt that the stag had been lying down for he suddenly sprang up within 100 yards of us and in the same instant the dogs were slipped. They had viewed him the moment he sprang up out of the heather. For a few seconds the stag took up the hill but the hounds ran cunning and cut him off; he now took a straight course along the face towards the direction where the carriages were waiting below. The hounds were going madly and were gaining on him. I now felt certain that he could not breast the hill at such a pace, therefore, instead of following over the rough ground we made all speed direct for the bottom to gain the main road. It did not take long and away we went as hard as we could along the road towards the direction of the carriages. As we drew near we could see the hunt. The deer had passed the spot where our party was in waiting, but he now turned down the hill towards the river with the two dogs within a few yards of his heels. Presently we lost sight of everything; we pushed forward past the carriages, and after running about a quarter of a mile down the road we heard the bay, and shortly arrived at the spot where the stag was standing in the middle of a rapid and the hounds were baying from the bank. No doubt the dogs expected to hear the crack of a rifle and to see the gallant stag totter and fall into the foaming river according to their old experiences. However they were not long in doubt. Patting both the excited hounds on the back and giving a loud halloo, I jumped into the water which was not more than hip-deep but the stream was very rapid. The stag on seeing my advance ran down the bed of the river and halted again after 50 or 60 yards. The two keepers had followed me, and Oscar and his companion no longer thought of baying from the bank but, being carried forward by the current together with ourselves, were met by the stag with lowered antlers. I never saw dogs behave better although one was, for a moment, beneath the water; Oscar was hanging onto an ear. I caught hold of a horn to assist the dog and at the same moment the other hound was holding by the throat. The knife had made its thrust behind the shoulder and the two gillies caught the horns fast to prevent the torrent from carrying away the dying animal. This had been a pretty course which did not last long but it was properly managed and in my opinion ten times better sport than shooting a deer at bay. I'm afraid that Sandy McCarra never quite forgave me for that hunt. 'Weel, you've just ruined the dogs for ever and there'll be no holding them from the deer noo. They'll just spoil the flesh and tear the deer to pieces.' This was the keeper's idea of what I thought was good sport. Certainly the venison did not belong to me nor did the dogs." [!]

CHAPTER TEN

The Lurcher and Longdog Overseas

The British Dominions and Colonies were founded by trade and military operations and it was many years before cattle and sheep were bred locally in sufficient numbers to provide food for the incomers. The early colonists relied on hunting to feed not only themselves, but also their native workmen and servants, with meat; and a great many greyhounds and deerhounds were exported to America, Canada, Australia and New Zealand, to help with the hunting. Whilst there are many accounts of such hunting, it is odd that so far as Africa is concerned, there is little or no history of coursing dogs; such great hunters as Selous and Cotton Oswell hardly mention dogs at all apart from the packs of kaffir dogs that guarded the camp at night and alternatively helped and hindered in following up wounded game.

Coursing dogs were used in India for sport from the earliest days of John Company, almost up to 1947 when the British left that hunter's paradise. Greyhounds and deerhounds were in great demand in the early days in Australia and from various cross-breedings the kangaroo hound was evolved. They are still used for hunting fox, pig and kangaroo. In America, hybrids of greyhound, deerhound, borzoi and saluki are used for hunting coyotes and red foxes.

A certain amount has been written about lurchers and longdogs overseas but it is a long way short of the whole story. As I have already quoted Sir Samuel Baker on coursing red deer, I continue with him, coursing sambur in Ceylon; I say something about coursing as I knew it in India and I quote from kind correspondents in America and Australia.

CEYLON

Sir Samuel Baker lived in Ceylon in the 1850s and did a great deal of hunting with a mixed pack. When he went out in the morning he did not know what the day's quarry might be: sambur, spotted deer, wild boar or leopard. Most of his descriptions are of hunting and coursing what he often alludes to as the "elk" but it is, in fact, the sambur (*Cervus unicolor*), a native of India and Ceylon. A stag stands about $13\frac{1}{2}$ hands and weighs from 550 to 600 lbs. The horns are not shed annually but irregularly every three or four years. Baker's descriptions of hunting sambur in Ceylon are probably the best ever written of the use of lurchers in what is, due to the march of 'civilization', no longer to be found anywhere except, perhaps in Australia – and that, illegally. Those who have tried to keep close to Fell hounds will realise how very fit Baker and his brothers must have been.

"No person" says Sir Samuel, "can thoroughly enjoy elk hunting who is not well accustomed to it, as it is a sport conducted entirely on foot, and the thinness of the air in this elevated region [over 6,000 feet] is very trying to the lungs in hard exercise. Thoroughly sound in wind and limb with no superfluous flesh must be the man who would follow the hounds in this wild country – through jungles, rivers, plains and deep ravines, sometimes from sunrise to sunset without tasting food since the previous evening with the exception of a cup of coffee and a piece of toast before starting. It is trying work but it is a noble sport: no weapon but the hunting knife; no certainty as to the character of the game that may be found; it may be either an elk, or a boar, or a leopard, and yet the knife and the good hounds are all that can be trusted in.

"It is difficult to form a pack for this sport which shall be perfect in all respects. Sometimes a splendid hound in character may be more like a butcher's dog in appearance but the pack cannot afford to part with him if he is really good. The casualties from leopard, boar and elk and lost dogs are so great that the pack is with difficulty kept up by breeding. It must be remembered that the place of a lost dog cannot be easily supplied in Ceylon [in the 1840s and '50s]. The pack now in the kennel is as near perfection as can be obtained for elk hunting, comprising ten couple, most of who are nearly thoroughbred foxhounds with a few couple of immense seizers, a cross between bloodhound and greyhound, and a couple of large, wire-haired lurchers, like the Scottish deerhound. In describing the sport I must be allowed to call up the spirits of a few heroes who are now dead, and to place them in the vacant places which they occupied in the pack.

"The first who answers to the magic call is Smut, hero of at least 400 deaths of elk and boar. He appears – the same well remembered form of strength, the sullen growl that greeted even his master, the numerous scars and seams on his body: his sire was a Manilla bloodhound which accounted for the extreme ferocity of the son. His courage was indomitable; he had several pitched battles with leopards from which he returned, frightfully torn but with his yellow hair bristled up and his head and stern erect; and his deep growl, with which he gave a dubious reception to both man and beast, was on such occasions doubly threatening. I never knew a dog that combined superlative valour with discretion in the degree exhibited by Smut. I have seen many dogs that would rush heedlessly onto a boar's tusks to certain destruction; but Smut would never seize until the proper time arrived and when the opportunity offered he never lost it. This rendered him of great value in these wild sports where the dog and his master are mutually dependent upon each other. There was nothing to fear if Smut was there; whether boar or buck, you might advance fearlessly to him with the knife, with the confidence that the dog would pin the animal the instant it turned to attack you.

"The next dog who claims a tribute to his memory is Killbuck. He was an Australian greyhound of the most extraordinary courage; he stood 28 inches at the shoulder with a girth of 31 inches. Instead of the surly and ferocious disposition of Smut he was the most gentle and affectionate creature. It was a splendid sight to witness the bounding spring of Killbuck as he pinned an elk at bay that no other dog could touch. When once started from the slips it was certain death to the animal he coursed, and even when out of view and the elk had taken to the jungle, I have seen the dog, with his nose to the ground, following on the scent at full speed like a foxhound. His unguided courage caused his death when in the prime of his life.

"Next in the chronicle of seizers appears Lena who is still alive, an Australian bitch of great size, courage and beauty, wire haired like a Scotch deerhound. Next Bran, a model of a greyhound; and Lucifer, combining the courage, speed and beauty of his parents, Bran and Lena."

Baker describes the country over which he hunted and gives a sample of the amount of game that he killed with his pack; the 'kill' usually being effected with his hunting knife or, occasionally, with a spear: between March 24th and June 29th, 1852, they killed 28 Sambur and four wild pig. It must be remembered that the pig he was hunting was the wild boar (*Sus scropha*) weighing up to 400 lbs, capable of outdistancing a horse for half a mile and, with his tushes and his immense strength, perfectly able to outface a tiger or leopard.

Baker's method of hunting Sambur was to pick up the morning 'drag'

which, in the case of the Sambur would be from a stream where the deer
would have drunk before making for the jungle and the day's sleep. The
scenting hounds would pick up the drag: "In the meantime where was
the noble stag? He was by this time standing somewhere high upon the
hill but usually at some distance from the crest. With a paunch full of
green food and a gallon or so of water taken when he quitted the river
bank, he had been disposing himself for sleep when his attention was
aroused by the excited voices of the hounds. If one could have seen him
he would have been standing with uplifted nose and well picked ears
listening to what was music to us but was a death knell to a deer. When
attentive to the distant voices, quite half a mile away, he little dreamt
that long legged mute hounds were far in advance upon the scent. Here
we see the advantage of the cross with greyhound and foxhound or
bloodhound. These dogs would follow by scent or sight but would never
open. Much faster than the rest of the pack, they went ahead and gained
a position close to the stag before he had an idea of any enemy. This
was absolutely necessary to ensure a quick solution of the hunt. If the
stag was not pressed to the utmost at the outset he would have plenty
of leisure to breast the mountain and reach the summit long before the
pack. In that case he would cross the ridge and descend the slope on the
other side. That would be a case indeed when the buckle of the belt
would be drawn tight to prepare for a long day's work; when once his
nose was turned downhill the sambur would never stop and he would
probably run for 10 miles or more into the depths of ravines where he
would probably be lost.

"If on the other hand, the lurchers were to interview him before the
arrival of the pack the result would be magnificent. For the first burst
the stag would make straight up the mountain side but the full paunch
of a night's feed would quickly tell and the hounds, running light, would
overhaul him and the stag must turn. Then he would come crashing
through the jungle, running obliquely downhill but the lurchers would
be at his heels and would force him straight down the steep incline
where he would have the speed. In the meantime, listening to the notes
of well-known hounds, I could tell with tolerable accuracy the position
of affairs. Hearing that the pack did not cross the mountain ridge I knew
that the stag had not been able to attain it; he would therefore, perforce,
be coming down.

"Judging the point at which he would be compelled to break, by the
appearance of the country, I would run ahead with the two seizers –

greyhounds or lurchers – ready to slip them the moment he appeared. I could now hear the pack in full cry coming down the hill; a few moments and the stag would appear at the edge of the jungle. He would pause a moment to assure himself of safety before venturing upon the open plain. Away he goes and at that moment I slip the greyhounds and the course begins. The stag knows nothing of these new enemies and he is not going at his maximum speed: they are. The greyhounds are closing on him as he nears the stream that runs through the centre of the plain. Suddenly he sees the dogs within 100 paces of him and the true race begins. They are too quick and are on either flank. One turns a somersault as a vigorous kick sends the dog flying but the other has him by one ear. The discomforted hound recovers and rushes up again; the other ear is pinned. Now the strength of a sambur stag is seen. He gallops forward with the two hounds hanging onto his ears. The ground is rough and covered with large stumps of course grass; against these obstacles the bodies of the dogs are swung with terrific force as the stag ploughs onwards but the good dogs never relax their hold. At length the stag trots – now slowly – now he walks. The dogs regain their feet and hold on like a vice.

"In the meantime the view halloo has been given the moment the greyhounds had been slipped. The well known sound, repeated twice or thrice, had been answered by the pack and every hound came thundering down straight for the cry. The lurchers that had been running mute would be within view, and tearing to the assistance of the nearly exhausted greyhounds. The knife would not be far away, and upon coming up, a thrust behind the shoulder would finish the career of the noble stag."

Baker had many of his dogs wounded and, on occasions killed by sambur. In his *Lays of the Deer Forest* (1848), Sobieski gives some good descriptions of how different dogs would tackle a deer. "The experienced greyhounds rarely run at the deer's neck but come up close by his flank, and shoot up at his throat, too close for the blow of his horns, and to effect this they will sometimes for several yards run by his haunch until they feel the favourable moment for making this launch at his neck. There are, however, dogs which have peculiar modes of attack. Thus some will sieze the deer by the fetlock, and one hound named Factor, a small but very fleet and highly couraged dog, was accustomed to make a spring over the deer's croup and fix himself upon the nape of his neck, when he never failed to bring him down."

E. P. Gee

Sambur; a young stag in India
(United Provinces)

INDIA

Blackbuck; a mature buck with a
very fine head

C. H. Stockley

The Longdog that marked the Sandgrouse; a saluki-Rampur cross

INDIA AND PERSIA

The author's saluki with a gazelle in Southern Persia, Boxing Day, 1942

INDIA

The sambur that Sir Samuel Baker hunted in Ceylon is also a native of India, though living in jungles where coursing would have been difficult if not impossible. I have never heard of anyone coursing them though it may have been tried in the past. But there was a lot of other sport to be had in India with longdogs. The actual dogs used varied from pure-bred greyhounds, salukis, and the occasional Afghan hound, to the local breeds such as the Rampur hound. And even the ragged-looking dogs used by Gypsies and Criminal Tribes were not to be despised. I had a good dog from some Banjaris that caught a lot of hares and jackal until he disappeared; I always suspected that he was stolen back by the tribe from whom I bought him; a practice not confined to the Orient. The Rampur hound, originating from the Native State of that name, was about 24 or 25 inches at the shoulder, rather 'on the leg' and with not much rib-cage, and with a heavier head than a greyhound. He was not terribly fast and had not much character on his own but used as a cross with a greyhound he produced a tough dog that would stand up to hard going and hot weather and would tackle a jackal on his own. There were also Australian kangaroo hounds to be found – I had two before the war – of which, more later.

Blackbuck
The fastest animal in the world may be the cheetah, but he is a sprinter. In India, he had to be taken in a pony cart or bullock cart to within slipping distance of the buck and he gave up after a few hundred yards if he had not killed by then. The blackbuck is almost as fast as the cheetah and, seemingly, can keep up his speed for ever. A great many sportsmen tried to course blackbuck with all sorts of dogs but very few were successful unless the odds were weighted against the buck.

The blackbuck is what his name implies; the adult male stands some 30 to 34 inches at the shoulder and weighs about 80 lbs. He is jet black on the head, back and side, white underneath and with a white ring round his eye. The young males and the females are brown with white markings. The male carries straight horns with a spiral twist; a good head is anything over 25 inches, measured straight, and the last record I heard of was $31\frac{1}{2}$ inches, very nearly the height of the buck itself. They normally lived in herds of up to 50 animals though in a few places they congregated in enormous numbers. I was once driving through Bikaner, a State famous for its game, when a herd of blackbuck started to cross over the track in front of my jeep, as was their habit. I very soon lost count of heads but the herd took 11 minutes to cross over the track and that is a very large number of buck, moving at a hand canter.

When coursing, the problem was to get close enough to the blackbuck to start the dogs off. The buck normally lived on open plains with very little cover and even when shooting, some form of stalking horse was often used to get within range. One method was to crouch in a bullock cart which was driven round the herd in a decreasing circle until one was within final stalking range. Another method was to sit behind a camel sowar – most uncomfortable – and drop off the camel at a suitable distance. A blackbuck had to be hit in the heart or spine or he would gallop for miles; even with a broken leg. On a flat plain, distance is difficult to judge, the heat haze makes the target shimmer like a mirage and, more often than not, a group of villagers would appear behind the buck just as the final aim was taken. I usually tried to have a pair of dogs fairly close behind so as not to lose a wounded buck but very few shikaris could handle them without showing themselves and scaring the buck out of range. I tried all sorts of schemes to slip dogs on terms with blackbuck; stalking, putting the dogs in a bullock cart, hiding a pair of dogs and driving the buck past them; but I have to admit that almost the only times I was successful was when coursing them on black cotton soil in the monsoon. The buck, with his small, sharp feet would sink into the soft earth and dogs could gallop across it without too much trouble.

Jackal

The jackal, Kipling's Tabaqi of *The Jungle Book* ("tabaqi": the plate-licker) was found throughout India and was the quarry of the various packs of 'fox' hounds such as the Peshawar Vale, the Poona and the Lahore Hounds. He also provided good sport when hunted with longdogs.

He lived on what he could pick up in the way of rodents, birds, beetles, etc., but was also a scavenger and was not above digging up the shallow graves in Muslim cemeteries after burials. Being a scavenger, he could often be found in the vicinity of villages or Cantonments (barracks) and could be met with in the early morning, returning to his earth from a village refuse heap. In the rains, earths would often be flooded out and the jackal would retreat to higher gound; in the middle of the day they could often be found lying up in crops such as a sugar cane or millet, from which they had to be driven. A native method of beating a strip of such crop was for two men to stretch a rope over the stalks and to walk slowly down the crop, one on either side of it, sawing the rope backwards and forwards between them. No game could stand such a noise over its head and out would come quail, black partridge, wild cat, porcupine perhaps and the jackal if he was home. I have, incidently, used this method successfully for beating a strip of clover for hares when short of beaters: some links of chain, added to the rope, bump along the ground and will move any hare no matter how tight it intends to sit.

Like a boar, a jackal had to be given time to take his line or he would dodge back into the crop. But once started he would run straight and one could have some good coursing if the ground was not too rough. On smooth ground, greyhounds would come quickly up to a jackal but unless they were very tough dogs they would be slow to tackle and would lay off a yard or two when they saw his teeth. The answer was a cross of Rampur hound or kangaroo hound. The offspring were not so fast as the pure-bred greyhound but were very much tougher, could stand the hard going and would usually tackle a jackal without hesitation. At one time I had a Staffordshire bull-terrier bitch that ran with the longdogs. Perhaps 'ran' is not quite the right word but she always caught up in the end and, of course, did not hesitate to tackle whatever was in front of her. I had acquired 'Judy' from a brother officer who was as impoverished as I was, and she had learnt to discriminate between natives entering one's compound bearing bills, and those who came in peace or with information about game; a most useful accomplishment in a country where bills were rendered by hand!

Even those people who did not see many jackals in their time in India, knew the jackal's cry. It was a long, wailing howl, repeated three or four times on a rising scale and ending with two or three sharp yaps. This cry would start just after dusk for about 15 minutes and would be repeated just before dawn. One could hear it approaching from a distance as jackal after jackal took it up and it would fade away again into the night. I have heard countrymen say that the cry started as the sun went down on the East coast and it followed the dark across India. Similarly with the dawn cry, heralding the sun from Calcutta to Karachi. It was a noise that was guaranteed to set every dog on edge and I am sure that many remember the first – and only – time one tied one's dogs to the tent pole in camp. At the first jackal's howl the dogs were off, whipping up the tent pole and leaving a collapsed tent behind them.

Fox and Hare

The Indian (desert) fox was very small, being about the size of an English hare, with enormous ears and a very large brush for its size. With this as a rudder it could turn even sharper than a hare and often beat the dogs who missed their strike time after time. I always regretted that I did not have a good pair of whippets to try out after the desert fox. Unlike the jackal who, if given some law, would run straight, the Indian fox ran in a circle if hunted by 'smell dogs' but the usual outcome when coursing was for the fox to go to ground. The Indian hare – again, smaller than the English hare – also had this trick and the look of surprise on the faces of two longdogs when the hare, that they were about to pick up, suddenly disappeared down a hole, was always comical.

I wonder how many blackbuck there are in India now? When the British left in 1947 there were estimated to be some 40,000 tigers, alive and roaming the jungles, The present figure appears to be about 2,000 and if the big game can be massacred in this fashion there must be little hope for small game. With the present publicity about the dangers of rabies, and all sorts of plans being spoken of such as a 'scorched-earth policy' if rabies is discovered in England, it is interesting to look back at the way in which rabies was virtually ignored in India. Any dog or fox or jackal behaving in an unusual way was shot as soon as possible but otherwise one never gave it a thought.

And One for the Miscellaneous Column of the Game Book

Whilst attached to the 13th of Foot, in Poona, before the war – a wonderful place in which to start one's soldiering – I took over a dog from a time-expired soldier in my Platoon, who was going home. The dog was reputedly saluki × Rampur hound. Having lived in a barrack room, he had been taught various tricks, one of which was to go down the side of the room either jumping from bed to bed or from space to space over the beds, to the word of command "beds" or "spaces".

There was a lot of sandgrouse shooting to be had in the Deccan and he quickly learnt his part in this sport. Sandgrouse live out on open plains and bare cultivation and are both hard to see and hard to approach in such conditions. But their habit is to fly in to water every morning between about 8 and 10 a.m., depending on the season. All the flocks in the neighbourhood would use the same watering place, usually a pool in an otherwise dry nullah. They would flight in flocks, settling for a while from half a mile to a mile out, and then, one after another, large packs of them would fly in, fast and straight, to drink. Although one knew to about quarter of an hour when they could be expected, they would come in from any direction and some prior warning was welcome. This longdog from the barrack room loved any form of shooting and soon learnt to connect the call of the sandgrouse with fast and furious action and, hopefully, some birds to find and retrieve. He would lie, apparently asleep; suddenly he would wake up, listen, and 'point' in the direction of the birds he had heard. His head would follow the sound if the flock was circling out of sight and he would 'lock on' as they started to come in to water, several seconds before I would see them. After some time he became so accurate that I could sit, reading, with my gun unloaded beside me and only at the last moment did I have to close my book, load, and face the direction from which the sandgrouse were approaching. It was a form of 'one-up-manship' that seldom failed to win a wager. Apart from his shooting exploits, this dog had plenty of heart and would tackle a jackal if he came up with him.

A Wartime Friendship

The dog shown on page 111 came from near Kasvin, in North West Persia, in 1942. I bought him for 100 cigarettes from the headman of the village near which we were camped for a while. The villagers said that he was about two years old and, so far as one could see, he was pure saluki with no pi-dog blood. As the photograph shows he was black with tan points and with very little feather; he stood about 24 inches at the shoulder or perhaps a bit more and when I first had him I doubt if he weighed much more than 24 lbs. He was skin, bone and sinew.

His first few days away from his village were not without incident; tethering rope was cut with one bite and signal cable did not take much longer to get through, so he had to be tied up with a chain, but he soon settled down in the Muslim Squadron of an Indian Cavalry Regiment. Urdu and Punjabi contain many Persian words and there were enough men in the Squadron, including my orderly, who spoke enough Persian to translate when an obvious misunderstanding occurred. The dog was used to feeding himself and though he was an artist at hunting anything that moved, from mice to marmosets, hares and foxes to unguarded sheep and goats, he was reluctant to hand over the spoils when they were suitable for human consumption. It took some time before he conceded that a hare should not necessarily be eaten on the spot, and one or two faint scars bear witness to long-ago arguments. In due course we moved down to Basra where he took part, with the many other salukis in the Division, in coursing gazelle. The photograph shows him with a gazelle he had pulled down after a five-mile course; his opponent had collapsed some time earlier. Shortly after this we returned to India, taking many of our dogs with us. The troopship had come out from home and still had a cellar on board: Guinness had not been seen in India since early in the war and a case or two got the dogs through disembarkation. At the first opportunity I collected an English greyhound which I had left with friends; when coursing hare or fox the greyhound left the saluki standing but after chinkara or blackbuck the greyhound's first rush was soon exhausted and the saluki's stamina was very obvious. The change of country was his undoing; he was not, of course, inoculated and he died of distemper in the hot weather of 1943. He was a very tough dog and on several proved occasions he had protected my belongings from thieves.

AMERICA

In the United States, longdogs were and still are used for hunting coyotes that prey on chickens, lambs, sickly sheep and, when they could get away with it, on calves. The coyote, a smaller cousin of the wolf, weighs 35 to 40 lbs, is fast, can 'stay' and is a hard fighter; he carries a fine pelt that is worth good money in late autumn or winter condition. Hunting coyotes, whether from horseback, from horse-drawn wagon or from jeep or truck, could be a most exciting sport. The apparently flat plains are cut up by wet or dry stream beds, often with vertical sides, stones and rocks, patches of sage and brush. In open country the coyote had to be stalked, using every fold in the ground, to get the dogs away on terms; then came the flat-out gallop across country, taking obstacles as they came. One writer who did a lot of this hunting between the wars gives his preference for dogs as: first, the pure-bred greyhound, then the 'cold blood', or not quite so pure greyhound; next the deerhound, the borzoi and the hybrid between them. And lastly the Irish Wolfhound, not for his speed, for he has none, but only if he was a fighter.

A correspondent from Minnesota writes: "The American houndman is very pragmatic and yet at the same time haphazard. He merely breeds one good hound to another regardless of background. Many simply buy in or trade hounds to better their packs. Having talked to men with 50

M. H. Salmon

Nebraska coyote hound; any rough-haired running dog is called a "staghound", as is also the case in Australia

U.S.A.

Nebraska coyote hound; 29½ inches at the shoulder and weighing 90 lbs. "The best all-round coursing hound I've ever known"

M. H. Salmon

M. H. Salmon

Texas; a "cold-blood" greyhound with two black-tailed jack-rabbits and a cotton-tail rabbit

U.S.A.

Nebraska; two lurchers and a saluki tackle a coyote. The coyote will be pinned down with the hay-fork and quickly despatched with a rap on the head

M. H. Salmon

years' experience of hounds I conclude that the basis of the American lurcher is the greyhound crossed first with the Scottish deerhound; secondarily with the Irish Wolfhound and Borzoi; rarely with the whippet or saluki. This breeding pattern may be explained by considering the game coursed in North America, the hare (jack rabbit), red fox and coyote.

"For hare, hunters here prefer the greyhound, usually what we call a 'cold blood' greyhound. These are pure-bred or nearly pure-bred greyhounds but they are not registered and come from a long line of coursing greyhounds. As opposed to track greyhounds (hot blood), they are long-winded, tough and sound, though slightly slower for the first half mile. Unlike the show greyhound, they are not over-size (usually 50 to 75 lbs). Occasionally, whippets and salukis are used for hares or are crossed with the greyhound but mostly it's the cold-blood greyhound for jack rabbits.

"The red fox requires courage as well as running ability from a hound. Again, the cold blood greyhound is preferred by many but large, mixed-breed, sight hounds, some showing the wire coat of the deerhound or wolfhound, are not uncommon in the mid-West where most fox are coursed. I'd say that any hound of 50 lbs or more can kill a fox singly, if courageous and determined. Foxes fight hard when cornered but are quite fragile (scarcely bigger than a big jack rabbit) once a good hound takes a grip.

"The coyote currently is hunted more than the fox or hare and here is where the breeding of lurchers is taken most seriously. Indeed, I believe a strain of hound is in the making, the American Coyote Hound.

"Among the pure-breds, only the greyhound sees much use today on coyote and mixed breeds are most common of all. Some are short coated, like the greyhound, others show a deerhound type of coat. Most are between 75 to 100 lbs; few hounds larger than this are good workers on coyote as they lack speed; the coyote is equal to the jack rabbit in speed and endurance but not in agility. In addition, the larger dogs tend to break down on rough terrain, especially in the fore-quarters.

"As regards the pure breeds, the deerhound and Borzoi are ideally suited for coyotes but so few are hunted that most individuals are found lacking, either in speed or desire. I don't believe a single kennel in America today is currently raising either deerhounds or Borzoi primarily for hunting. As as for the Irish Wolfhound, I've yet to see one that was more than one stride better than useless on any game.

"For me the ideal coyote hound would stand 28 to 30 inches at the withers and weight 80 to 90 lbs. He would have a deerhound-type coat, as our prairie winters are severe, and as protection from barbed wire and from cuts from fighting coyote. His speed would nearly equal the field greyhound and he would be able to take game at two miles if necessary. As a pair, he would kill a 40 lbs coyote unaided. I have seen such hounds in Nebraska so I know it's possible. Few Afghan hounds or whippets are used or bred for hunting in America. I've only seen one Afghan I'd care to own and just a few whippets. Many whippets are raced and are fast but they lack the stamina to catch the jack rabbit. They are, however, excellent on the cottontail rabbit."

AUSTRALIA

It is interesting to read about the size of dog preferred in America, 80 to 90 lbs and 28 to 30 inches at the shoulder; a size necessary for the great open plains but too big for the much smaller scale of the English countryside. In Australia, with vast areas of open country and very rough bush, the same demand for the bigger dog resulted in the kangaroo hound.

The early settlers in Australia found game in plenty and, to help them in hunting, they imported English greyhounds. These were found not to be big enough to tackle kangaroos, particularly the big red kangaroo, standing over seven foot high, nor could the greyhound stand up to the local conditions: hard ground, sun cracks, fissures, thick undergrowth, and the trees, living, dead and burnt. Various crosses were tried to mastiff, Dane, boar hound, bloodhound, to name a few, but the most successful was the cross with the deerhound and perhaps a touch of cattle dog added. This produced a dog of about 30 inches in height, very muscular, and weighing upwards of 75 lbs. In more open country and where the smaller, grey kangaroo was common, a cross back to the greyhound was made for greater speed. As cattle and sheep increased so the demand for kangaroo meat lessened and the kangaroo hound started to lose his vocation. Then came the demand for kangaroo hides for leather work and for kangaroo meat for the pet-food trade and the kangaroo hound came back into his own – though in reduced numbers – mainly to help with hunting at night with rifle and spot lamp.

I had two kangaroo hounds in India before the war; big dogs, rather ugly, with heavier heads and jaws than the greyhound, one almost black and the other a dark brindle. I believe they had originally arrived in

A Red Kangaroo, standing over seven
foot high

AUSTRALIA

Deerhound and foxhound with a
Grey Kangaroo

A successful morning with a mixed pack of deerhounds, borzois and rough-haired lurchers

AUSTRALIA

Chas. Venables

India on a boat bringing Walers (Australian horses for the Army) to Calcutta. But as I have recently read an Australian book on greyhounds in which it was stated that coursing is almost finished in England, the only Clubs left being Altcar and Swaffham (I wonder what had happened to the other 20 Clubs?), I must be careful what I say about longdogs in Australia. I am therefore all the more grateful to the many kind correspondents who have brought me up-to-date on kangaroo hounds.

The kangaroo's methods of defence are simple; when attacked, he backs up against a tree or a rock so that his rear is guarded. If a dog jumps for his throat the kangaroo catches him with his fore paws and, balancing on his tail and one hind leg, he lifts the other leg with its spiked toe and with one kick he disembowels the dog from throat to stomach. So will he treat any other dogs foolish enough to rush in. But dogs that know their work will not run in to be killed; one of them feints at the kangaroo in front whilst the other tries to grab his tail and upset him. If water is nearby, the 'roo will make for it and plunge in, waiting for the dog to swim to him. Here again, the headstrong dog is seized and drowned, unless the other can grip the 'old man' and pull him over backwards to release his partner. Because of the kangaroo's habit of kicking up with a hind toe, many kangaroo hounds wore a broad leather collar to protect their throats at close quarters.

As might be expected, the breeding of kangaroo hounds varies with the breeder concerned. Although some are still used for kangaroo – more often than not in following up 'roos wounded at night when spotlamping – many are used for hunting pig and also for fox. One correspondent who crosses greyhounds with bull terriers says: "We usually go away every three weeks or so; we hunt pig in the morning and afternoon and then we are out of a night with a couple of dogs in the back of the ute spotlighting for 'roos." He encloses a drawing of a dog wearing what looks like a five-inch-wide leather collar for protection. Many writers use phrases such as "this grand old breed of dog that is disappearing faster than the kangaroo", and several mention how rare it is to see a kangaroo hound advertised nowadays. One says: "Another problem in definition and description is that the Australian bushman only talks about two types of hunting dog, the staghound and the beagle. The staghound is any rough greyhound and the beagle is any dog that bays on the scent."

"My present hound is a double cross of greyhound" writes one hunter;

"He was bred mainly for catching foxes and this he does more than adequately, but he also has a great time on 'roos. About three weeks ago a station owner contacted me to come and thin out his 'roo population; my hound successfully accounted for 14 of these furry pests in one night. Whilst hunting recently in bush country, three dogs put up a big 'roo and brought him to a standstill, but being over seven foot tall he was a bit hard to bring down. However, one dog sneaked in and bit this old buck in the scrotum, thus gaining his undivided attention; with this opportunity the others seized the 'roo and finished the chase quickly."

A veteran gives an interesting comment on conditions in a small community just after the First World War: "I came to Australia direct from the British Army in 1920 and was sent out to newly settled land, where lots were given to ex-Servicemen. Conditions were primitive; and it was soon apparent to me that there were only two main things by which a man was judged as to his standing in the community and they were (1) if he was of Catholic religion and (2) if he had a good dog. The religious aspect was pursued with great intensity and bitterness but strangely enough the Protestant could and would mix with the Catholic if either owned a good hunting dog." I wonder if there is a lesson for Ulster here.

Another veteran with 50 years' experience of hunting in Australia found that though cross-bred dogs were just as good, and often tougher than pure breeds, "I could get $40 for a cross pup as against $100–150 for a pure. We hunt mainly 'roos, emus, pigs and foxes. Landowners contact me to ease their pest problems; 50 to 100 'roos or emus in a grain crop can create Hell. I have to travel about 25 miles to reach hunting territory so I use my Land Rover. Fifteen years ago we did it all on horses but hunting 'roos and emus is illegal so you are sticking your neck out to set off with a horse float and hounds. I like a medium-sized hound, about 31–33 inches. Rough-coated dogs stand up to knocks better than the smooth; bitches mature quicker and are less quarrelsome in the yards.

"I have tried greyhounds but they are no good in rough country, too easily hurt. Deerhounds are the tops; gave up borzois after five years, too temperamental (like the Russians). Wolfhounds can be very game but are not fast enough. Afghans are a write-off, too much coat and no brains. Salukis are a lovely hound, fast and tough but they are the only breed of dog I could not control, could not get 'through' to them."

His "medium-sized dog of 31–33 inches" would find the cramped conditions in England not much to his liking.

CHAPTER ELEVEN

The Lurcher and the Law

I make no pretence of trying to interpret the Game Acts for the benefit of lurcher-owners; for one thing, the law is so complicated that few laymen can find their way through it with confidence; and for another, few lurcher-owners have even heard of the Game Acts and the remainder couldn't care less about them. And, as I shall show, this attitude is not exactly new; but it is interesting to see how the longdog has been legislated against for at least 960 years.

The Forest Charter of 1016 stated that: "No mean person may keep any greyhounds: but freemen may keep greyhounds so that their knees be cut before the Verderers of the Forest and without cutting of their knees also if they do abide ten miles from the bounds of the Forest. But if they do come any nearer the Forest they shall pay twelve pence for every mile: but if the greyhounds be found within the forest the master or the owner of the dog shall forfeit the dog and ten shillings to the King." There were two dogs that might be kept by those who legally lived within the Royal Forests; firstly the mastiff. The times were lawless, criminals and outlaws survived by their strength and their wits and everyone living or having property inside the forest bounds might keep a mastiff to protect themselves and their property. But the mastiff had to be expeditated or 'hombled' to make sure that he could not chase the King's deer. Expeditating was the chopping off of one or more toes of a front foot, usually the right-hand one. Occasionally the tendon behind the knee was cut so that the dog could not gallop but cutting off the toes was the more usual method of crippling. The other dog that could be kept, and which was safe from mutilation, was one so small that it could pass through a stirrup measuring five inches by seven inches; another test was that the dog should pass through a strap, "eighteen inches and a barleycorn in length".

As I have already shown in Chapter One, the Norman Kings had appropriated as much as a third of the whole country as Royal Forests, firstly to provide game and revenue for the Court and secondly for the King's own hunting. Over and above the Royal Forests, the great landowners had their chases, parks and right of warren, consequently there was little land left where the public could hunt. The diet of the ordinary villager was monotonous at most times and in the winter often came near to starvation point. Hunting, whether it was with bow and arrow, crossbow, snare, net or dog, was therefore not a matter of sport – this was for the rich – but of necessity. Virtually the only places where the common people could hunt with any hope of finding game were the Royal Forests and all through the Middle Ages the Forest Courts were kept busy, investigating cases of poaching and dealing out punishments. The stories told in court by the accused sound very modern at times: "That Ralph of Halcot, a clerk, was journeying towards Huntingdon with a page who led greyhounds. And they escaped from his hands and brought down a fawn of one year old." And again: "Not but that I will confess that my two greyhounds escaped from the hand of my small boy by reason of his weakness, and that I followed them to the park and entered there by a breach that I found already used, and pursued my hounds and retook them". Putting the blame on someone else is not new.

Poachers were not above cocking a snook at authority: "That the said William and others killed three deer without warrant and they cut off the head of a buck and put it on a stake in the middle of a certain clearing, placing in the mouth of the aforesaid head a spindle; and they made the mouth to gape towards the sun in great contempt of the lord king and of his foresters."

In King John's time, poaching became so rife that he ordered all dogs kept in forests to be slaughtered. Sunday lurcher meetings are no new thing, as witness a statute of Richard II, of 1390: "For as much as divers Artificers, Labourers and servants and grooms keep greyhounds and other dogs and on the Holy Days when good Christian People do be at Church hearing Divine Service, they hunting in Parks, Warrens and conigries of Lords and others to the very great destruction of game ... it is ordered that no manner of Artificer, Labourer nor any other layman which hath not lands or tenements to the value of XLs [£10] by year, nor any Priest nor other Clerk if he be not advanced to the value of XL, shall have or keep from henceforth any greyhound to hunt; nor shall they keep Fyrets, Heys, Nets, hare pipes nor cords." In 1539 there were complaints in Parliament against the new, lay, owners of the monastries:

". . . the unreasonable number of hounds and greyhounds which the gentlemen keep·and compell their tenants to keep, and many tenants do keep them for their own pleasure. These dogs eat up the broken meats and bread which should relieve the poor. They say they must keep dogs or their lambs would be killed by the foxes".

James I signed "An Act for the better preservation of Deare, hares and other Game aforesaid" which laid down fines for those who kept greyhounds and who had not property to the value of £10 a year or more; a fine of 40 shillings imposed on a man who probably did not handle any money from one year's end to another could have meant imprisonment for life. Charles II's Act of 1670 is the first one to mention the lurcher by name: "Wheras divers disorderly persons laying aside their lawful trades and imployments doe betake themselves to the taking, stealing and killing of conies, Hares and other Game", authorises gamekeepers to sieze "all such Bowes, Gunns, Greyhounds, setting-dogs, *lurchers* and other dogs to kill hares". Queen Anne's statute on similar lines went further, in providing case law to show that even the keeping of a lurcher was an offence in itself, and it only needed the oath of one witness for a conviction. An interesting statute was passed by Parliament under Henry VIII which said that: "No person of whatever estate or degree shall trace or kill any hare in the snow." Under George III, in 1798, a lurcher was rated with "greyhd, hound, pointer, setting-dog, spaniel or terrier" at six shillings per annum.

In the early 19th century the dog tax was levied on all dogs except those that had a short tail, these being the bobtail sheepdog, on the principle that no one would keep a mutilated dog for other than work. Many lurchers of the time had their tails docked in the hope that, since most of them looked like sheepdogs they would neither be taxed nor shot on sight. By 1850 the tax included all dogs whether docked or not and the docking of sheepdogs stopped.

However, to the Game Laws. The old fashioned 'keeper solved the problem of complicated laws by shooting a lurcher on sight; happily this is no longer a solution and a dog may not be shot when in pursuit of game as there is no property in game until it is 'reduced into hand'. Pheasants in a rearing pen are 'in hand' and are the property of the landowner or whoever put them there. Some weeks later, the same birds in covert are no-one's property until shot or otherwise secured. It is different if a dog is chasing farm stock; it may then be shot, but only if there is no other possible method of stopping it doing actual damage.

Hares, being included in the category of Game, may not be taken – by any means – on a Sunday or Christmas Day; how many lurcher-owners have refused an invitation for Sunday coursing on this score? A game licence is necessary if hares are taken by any means other than with hare hounds or at a coursing meeting, though the meeting does not, apparently, have to be a properly organised one. I suspect that many of us break at least one law on many Sundays between September and March.

If you, the lurcher-owner, "commit any trespass (entry without prior permission of the occupier) by entering or being (physical presence) in the day time (one hour before sunrise to one hour after sunset) upon any land in search of or in pursuit of game (hares) or conies (rabbits) you commit an offence"; in other words you are poaching. The Game Act 1831, as amended, says so and don't be misled by the date of the Act. You may note the let-out in that physical presence is necessary; sending a dog into a field is not enough to commit an offence and there is, apparently, case law to support this. "So", you say, "I'll set a net across this gate by the lane, I'll send the dog round the field and I'll wait here in the ditch which belongs to the Council and I'm quite sure they are sitting in front of their tellies at this time of the evening". So far as I can find out, the answer is that 'land' includes any public road, highway, path, the sides of them and the openings and gates onto them; and taking hares or rabbits on any land with a "net, engine, or other instrument for the purpose of destroying game" means that you can be arrested in your kitchen if that is where they have chased you to. And if, whilst being arrested you offer violence to one of the many people, besides the police, who are authorised to arrest you, you may be liable to seven years inside. Not all that long ago it would have been 14 years' transportation!

It is interesting that dogs may be seized when chasing deer which are enclosed – in a park – but not when chasing deer that are roaming the countryside. The two latest Deer Acts (Scotland, 1959, and the Deer Act 1963) make no specific mention of dogs; but the Scottish Act makes it an offence to take or wilfully kill deer on any land without legal right or the permission of a person having such right, and also makes it an offence to take deer otherwise than by shooting with a firearm. The only exception is the prevention of suffering. The Deer Act (1963) makes it an offence to take.deer during the prescribed close seasons or to take deer at night (one hour after sunset to one hour before sunrise). It also makes it an offence to kill deer by traps, snares or poison; or to use a firearm of under a certain calibre; or to use arrows, missiles or spears; or to discharge anything at deer from a car. To my layman's eyes there is nothing in the Act that makes it an offence to take unenclosed deer with dogs in the daytime and outside the close season but I am most probably wrong and I take care not to have to prove my theory.

It all boils down to the fact that if you haven't got permission to be where you are then get your business done as quickly and quietly as possible and get away.

CHAPTER TWELVE

Lurcher Shows and Judging

Lurchers are spectacular dogs and even a small lurcher show will draw a crowd if it is publicised; such shows are fun so long as they are treated light-heartedly, and some racing or carnival events or show jumping can be added. Most lurchers will jump considerable heights and many of them will jump for the fun of it. I offer some suggestions for running a lurcher show; they have been tried out on several occasions and seem to work well.

Men and Material
1. "Men"
   a. Two – with one relief – so three in all, to take entries; one takes the dog entries and the other takes the bitch entries. This saves a lot of time particularly at the beginning when there is usually a rush. If the entry-takers are pretty girls with flashing smiles the work will go the quicker and there will be fewer arguments.
   b. Two Ring Stewards; tactful men who can make people get a move on; if no loudspeaker is available they should have loud voices.
   c. One Judge or two Judges. From the Judge's point of view it is easier and quicker to do it alone but more amusing and interesting to judge with someone else.
   d. A Referee in case the Judges can't agree.

2. Material
   a. A ring, at least 20 to 30 yards by 20 to 30 yards, roped off, or, better still, surrounded by netting such as electric sheep netting. An entrance at the corner nearest where the entries are taken.
   b. Straw bales round the ring for spectators to sit on.

    c.   A side-loading horse-box – or open-fronted tent – for the people taking entries.

    d.   Tables, chairs, mil-boards with paper clips, lined paper, carbon paper, biros, cash box for change and a megaphone.

    e.   Two sheets of hardboard or plywood at least 2 feet by 3 feet, painted white. On one is written the schedule of classes and on the other is written the names and addresses of the winners (see below, para. 10). I suggest writing with a broad, felt-tipped pen; people must be able to read the schedule of classes from some yards away. This means 2-inch letters at the least.

    f.   A piece of 5-ply, thick hardboard, an old door or some other slab about 6 feet by 4 feet, in the middle of the ring for the dogs to stand on while the Judges look at their feet. Grass can hide a multitude of sins.

3. Advertisement

This will probably be done by the organisers of the Show but there are various mailing lists of lurcher people available.

4. Classes

The number of classes rather depends on the time available. You can't start too early or no-one will be there to show their dogs; and you should stop for lunch at a reasonable hour. Unlike formal dog shows, lurcher-owners come for a day's amusement and will gravitate towards the beer tent soon after mid-day if not earlier. So, for practical purposes you have from about 10 to 12.30 or 12.45 into which the classes must be fitted. As a guide I suggest:

    1.   Rough Dog
    2.   Rough Bitch
    3.   Smooth Dog
    4.   Smooth Bitch
    5.   Puppies, under 12 months
    6.   Veterans
    7.   Pairs, or 'Families', i.e., Pa, Ma and three from one litter
    8.   Championship (winners of classes 1 to 6)

5. If there are a lot of entries the seven classes and the Championship will take some pushing through in 2½ hours. Puppies and Veterans could be left out if necessary. On the other hand, instead of these

A Class lined-up

LURCHER SHOWS

A well-matched pair

Mr Roddy Armytage and the author judging at Lambourn

LURCHER SHOWS

The author takes a close look at the Champion

two classes, classes for Rough Dogs and Bitches under 22 inches and Smooth Dogs and Bitches under 22 inches (Dogs and Bitches to be shown together to save extra classes) can be substituted if there is a lot of demand for them. Smaller dogs don't stand much chance against the full-sized ones of 25 and 26 inches.

6. Rosettes

Three rosettes for each class: 1st (red), 2nd (blue) and 3rd (yellow). In addition somewhat flashier ones for the Champion (red, white and blue) and Reserve Champion (red and white). Rosettes should have the name of the Show and the year printed on them. If any prize money is offered it sould be put into envelopes and the envelopes stapled to their appropriate rosettes before the Show starts.

7. Entry Fees

It is difficult to advise any sum as the Show finances will probably dictate such things. But a lurcher show should be fun and not an expensive and rule-ridden bore; I suggest 20p per dog as entry fees and £1, 50p and 25p as prizes. Whatever the sum fixed for entry it should be a round figure to 10p so as no 5p pieces have to be kept as change.

Operations
8. Entries

The people taking the entries should be seated in the horse-box at least half an hour before the first Class is due to start. Let us say it starts at 10.00; the girls should be ready in the horse box at 9.30 with their paper prepared for the various Classes, i.e., top sheet, carbon and copy sheet clipped together for each Class. One takes the Dog entries and the other takes the Bitch entries. Later on during the morning, when the rush is over, one person can probably do both but to start with it will save a lot of time if it is done in this way. At 9.30 there should be an announcement over the public address system to the effect that the "first Class, Rough Dogs, starts at 10.00 and entries can now be made for all Classes". Another shout at 9.45 and a final one at 9.55: "Still time to enter for Rough Dogs if you run!" All the girls need is the name of the owner, the name of his dog and his 20p per dog.

9. The First Class
    At 10 o'clock one of the Ring Stewards collects the top sheet of
    entries for the First Class and takes it across to the other Ring
    Steward at the entrance to the ring. One of them shouts
    (megaphone) "Class One, Rough Dogs into the ring now please",
    and competitors are checked off against the list as they come in and
    are asked to walk around the ring one behind the other. One of the
    Ring Stewards should keep an eye out for the customer who jumps
    his dog over the netting and joins in without having paid his entry
    fee.

10. Judging
    The Judge, or Judges, stand in the middle of the ring and they
    should let the competitors do at least four or five circuits before
    calling anyone in so that they all get their money's worth. Then
    they can start weeding out and it is better in a large ring if they
    tell the Ring Steward which dogs they want and the Steward calls
    them in: "That grey dog, the chap in plus fours, the girl in blue,
    the red brindle, the small boy with the fawn, the gypsy with the
    black", etc., etc. The ones called come into the middle of the ring
    and the remainder keep walking on round until the Judges have all
    they want; seven or eight out of a Class of 25 to 30 are enough for
    the final selection. The Ring Steward then tells the rest they can go
    and "Thank you very much" or words to that effect, and the
    selected dogs are walked round again, once or twice. This time the
    Judges ought to call them in in the order of preference, and when
    they are lined up each one is asked to trot out and back, and then
    stood on the hard slab to see what their feet are like. When all have
    been looked at the Judges say "That's it" or they re-arrange the
    order as required. Meanwhile one of the Ring Stewards should have
    fetched the rosettes from the horse-box and as soon as the Judges
    are satisfied he should give out the rosettes; the other Steward
    should ask the winners their names, the names of the dogs, and
    their addresses. Quite often spectators see a dog they would like to
    use as a sire, or a bitch from whom they would like a puppy; once
    the winners have left the ring they are very difficult to identify in
    the crowd, but if their names and addresses are written up on the
    board on the horse-box, those interested can get in touch with the
    owners later on.

Peter Goulding

"Ask how they're bred. . .

LURCHER SHOWS (Holkham)

". . . and see how they move"

John Compton

John Tarlton

The Show is a day out for the family

LURCHER SHOWS

Lunch break

Author

11.　And so it goes on. Regular announcements over the P.A. system will be needed to keep entries coming in and the winners of each Class can also be given out. A quick gin and tonic for the Judges and Stewards about mid-day will keep a sparkle about the Show and as soon as the last Class is finished there is a call for the winners of each Class (less the Pairs Class) to come into the ring for the Championship. The Judges should select a Champion and a Reserve Champion. One of the Stewards should keep an eye on the time and warn the Judges if they are going too slowly. Slow judging bores everyone, but on the other hand if the judging is done too quickly the competitors will complain that their dogs haven't been looked at properly.

Judging

If there is to be a lurcher show then there must be a Judge to put the competitors in the right order; or, at least what he or she thinks is the right order. I know very well that the only proper test for a lurcher is to take the dog out to the countryside and put up a hare. But if one is faced with a class of up to 50, rough or smooth, dogs or bitches, one has to sort them out on their looks alone and, for what it is worth, I give some views on the subject.

The lurcher Judge has one or two advantages over the Judge at a 'pedigree' show; firstly there is no set precedence for judging lurchers so the Judge can use whatever method he thinks fit. The dogs which the Judge likes most can be pulled into the middle, whilst the rest are walking round, and then put in order; this is probably the easiest when the class is a small one. Or the ones which the Judge likes least can be thrown out until seven or eight are left for sorting into order. This clears the air when the class is very big. Another advantage the lurcher Judge has is that, unlike a formal show, where all the entrants will, or should, look like, say, whippets, a lurcher show is a free-for-all and there are certain to be some dogs in the ring that are not lurchers or do not look like lurchers. The third advantage is that, whatever the final placings, the results will not make any difference to human or canine reputations; the only lurcher reputations worth anything are won in the coursing field. The only person who cannot take the day too light-heartedly is the Judge.

Now, as to the actual judging. One hopes that the ring is big enough to allow a large class of large dogs to keep walking round; some may immediately catch the eye; so much the better as the rest can be looked at against them. Five or six circuits of the ring should be enough to look at every dog; a good way of doing this is to pick one point on the circuit and watch each dog as it passes that point, as if they are crossing a gap in a wall. In a large class I would then throw out anything that had not got a proportion of greyhound or deerhound in it; this would remove those that look like sheepdogs, alsatians, salukis, labradors and plain mongrels and it can reduce the class by up to a quarter. Next, I would throw out anything that did not look both functional and lethal. A lurcher may not have to live rough but at least it should look as if it was able to do so. The rough-coated classes pass this test but in the smooth classes I would look for a double coat and pass over any dog with a fine, silky coat like a whippet. I would then ask myself which of the dogs that are left look as if they could keep me in grub if I was living rough. A lurcher must be able to catch a hare – under most conditions – should be big enough to hold a deer, and not so big that it can not get down to a rabbit. This reduces the field still further to those dogs between about 23 and 27 inches.

Now comes the time to look more closely at conformation. I know that a lot of dogs that look like nothing on earth do, in fact, course very well; but how much better would they be if they were properly put together? The things I look for are:

Head:	long and lean, wide between the ears, a powerful jaw and a bold eye. The colour of the eye is immaterial but a shifty eye usually means a shifty dog.
Neck:	long and muscular, flowing back into the top of the shoulder. The top line of the neck should be longer than the under line.
Shoulder:	well laid back and the humerus bone nearly vertical.
Chest:	deep and wide, a handspan between the forlegs.
Ribs:	well sprung.
Back:	long but not too long, with a roll of muscle on either side of the spine.
Quarters:	sloping, with very muscular thighs.
Stifle:	well bent with a muscular second thigh; hocks close to the ground.
Bone:	not too heavy but enough for the dog's size. The bone should be flat and must continue below the knee.

Jim Meads

"Under starter's orders"
LURCHER SHOWS

"They're off!"

Jim Meads

Jim Meads

Lurchers running;

LURCHER SHOWS

Lurchers resting

Jim Meads

Feet: strong and compact, well-set-up toes, and short, trimmed,
               nails.

Since the lurcher is a working dog or he is nothing, I would penalise
a dog that had bad feet. One sees all too many of them: flat feet:
spreading toes and nails that are far too long. The dog that has one or
two toes let down, through hard work and the back tendon rupturing,
is the victim of his occupational hazards and this need not be treated as
a fault.

The other points such as colour of coat and length of hair that are so
important at dog shows are immaterial to the lurcher. But there is
something in the saying that a tawny eye means a hard dog; and I must
admit to a dislike of white dogs. I think they tend to be soft and there
is an old coursing predjudice against them on the argument that the hare
turns from a white dog quicker than from a darker one. Temperament
is not easy to assess but I like to see a dog that is alert and is watching
what is going on; and it is difficult to resist the dog that grins at one and
whirls his tail round when the previous two have nearly taken off one's
hand.

Movement is not easy to assess, as few lurchers walk to show themselves off; and when the owners are asked to trot their dogs "up to that
corner and back" the dogs think they are getting away from a boring
performance and tend to bound all over the place. But at a walk I like
to see a dog carry his head low and his action should be powerful and
forward-going. There must be a real drive from the stifle and the front
feet should be kept low to the ground. I don't mind a dog that goes a
bit close behind, particularly at the trot, but he must not turn his hocks
outwards. Similarly front feet that turn outwards should be penalised;
the dog usually has tied-in elbows.

Some Judging Problems
With the current explosion of interest in lurchers — which may or may
not be a passing craze — the Judge at anything but the smallest Show
may be faced with two problems: these are firstly, enormous classes to be
looked at in too short a time, and secondly, being asked to award an
overall Championship between the large and small lurchers.

Unlike a normal dog show a lurcher show is usually divided into a
morning of showing and an afternoon of racing and other fun and
games. The judge therefore has about three hours, at the most, to deal

with up to 9 classes. The first four classes can total 200 dogs or more and there can be up to another 100 entries in the additional classes, puppies, pairs etc; though many of these dogs will already have been looked at they must be looked at again in their new setting. Simple arithmetic shows that in this extreme case the judge has 36 seconds in which to look at each entry, assuming that they pass before him in an unending stream. But of course they don't; there are intervals whilst classes change, pauses while stewards fetch rosettes and other delays which cut down the actual judging time. Obviously something has to be done to reduce the size of each class very quickly to numbers that the judge can concentrate on; and my suggestions on pages 139-143 should be sufficient to do this provided that the judge carries them out quite ruthlessly. It is for this reason that I, personally, prefer to judge on my own; judging with someone else is usually interesting but it does mean time taken up in consultation and, inevitably, compromise.

The second problem, that of being asked to award an overall Championship, is more difficult to deal with as it has usually been arranged by the organisers long before the judge is appointed. I know that at normal dog shows judges of Any Variety classes are happy to differentiate between Great Danes and Yorkshire terriers; there is a Kennel Club "standard" for each breed. But there is no "standard" for lurchers, other than work; the judge must pick out the dog or bitch which looks most likely to "hunt, jump, catch, kill and carry" or, in other words, looks most likely to keep him in grub if he was living rough. With this as a "standard" there can be no competition between the 21 inch rabbiting dog — however nice it looks — and the 25 or 26 inch dog that can cope with deer as easily as it will catch hares and rabbits; the rosette *must* go to the big dog unless there is something very wrong with him and if there is then how did he — or she — win the large lurcher class? *In my opinion* show organisers should not ask the judge to award an overall Championship between the large and small winners; they are different dogs for different purposes.

CHAPTER THIRTEEN

Whelping and Rearing

A Strong Family Line

As I said in Chapter 4, most lurchers are bred from lurchers; this can result in absolute rubbish but the greyhound-collie cross can, and does, breed true to type if each mating is very carefully considered and no exotic breeds are introduced. From time to time some fresh greyhound blood may have to be introduced to keep up speed and perhaps there may be the odd touch of deerhound for rough, weatherproof coat and stamina though this outcross does usually increase the size; if smaller lurchers are required the outcross should be to collie.

As an example of lurcher to lurcher breeding, with an occasional outcross, I have included here a "pedigree" with photos of some of the dogs and bitches involved; these show how a strong likeness can run in a family, particularly in the female line. The line goes back through Cumberland and Northumberland to Norfolk, to the original greyhound-collie cross, reputedly a stolen service by Mick the Miller! A well-known greyhound trainer who knew these dogs at Kings Lynn before and during the War told me that they were "Big, shaggy Norfolk lurchers, most of them honey-coloured. Great big dogs, up to here; you don't see them like that now-a-days. No hare could live in front of them and they spent half their time poaching on Sandringham estates".

Whelping

The mechanics of breeding, mating, etc., have been described in so many dog books, that there is no need for detailed accounts here: I will merely mention a few stray thoughts on the subject.

If a bitch is in whelp she will quite likely to go off her food on the 21st day after mating and sometimes continues to pick and choose until about the fifth week. Thereafter she should be done as well as possible; steady exercise every day and as much meat as she will eat, milk if she will drink it and some calcium added to her food. I do not believe in a lot of additives; if a dog is being fed properly and has access to grass and certain weeds there should be no need for bottles and jars of expensive supplements. But for a large bitch that is going to produce up to nine or ten puppies, some extra calcium is essential; and though many people laugh at herbal remedies, I believe that a daily dose of raspberry-leaf pills does help cleansing at whelping time.

Most books tell the reader that, when whelping starts, the bitch should be left alone to get on with her job which she will do better without human assistance or presence. There are many stories of bitches who disappear when whelping time draws near and reappear some time later from underneath the chicken house or some such refuge, having produced a large and healthy litter. I remember a greyhound bitch of my father's, at Killaloe, that dug herself into a sandbank at the back of the farm and produced eight puppies out of sight and out of reach; when grown up they all turned out to be track winners in Ireland or in England. Of course most whelping is trouble-free otherwise the race of dogs would have died out by now, but I offer three reasons why the owner should be present during whelping; not necessarily present all the time but at least to try and see each puppy arrive. Firstly, one needs to know if all the after-births have come away; otherwise there will almost certainly be trouble. Secondly, if any complication should arise, one needs to know what stage the bitch has got to; if a puppy has not appeared after an hour and a half's straining, professional assistance will very likely be needed. If the bitch had been on her own until trouble is found, no one will know how long the trouble has been brewing.

The third reason is one that I never seen mentioned in print but which, to my mind, can be important when the question of which puppy to keep comes up. Unlike greyhounds, where whole litters are – or were – run-on to see which is the best, I doubt if any lurcher breeder can afford to keep a whole litter to maturity. Lurchers grow very unevenly, and choosing the puppy to keep can be extremely difficult; but I have found that some longdog puppies will show their later shape for a few moments when just born, whilst they are still wet and before the bitch has licked and tumbled them dry. It does not always happen by any means; one may not see it at all in a litter. But length and symmetry can sometimes be there for a moment or two. The first two or three puppies may have had nothing particular about them and then the fourth one arrives and as it struggles free from its caul one suddenly says to oneself: "That one is going to gallop." Within minutes the 'vision' has

"What goes in . . .   *Author*

". . . must come out"   *Author*

Jim Meads

"GYPSY", taking after her father, "DICK". A later photo on page 84 shows her at 18 months of age, well into her stride

LITTER SISTERS AT 6 MONTHS OF AGE

"DINAH", already standing with her hind feet under her like her mother and grand-mother, "TARN" and "SAKER"

Jim Meads

"GYPSY'S" and "DINAH'S" pedigree

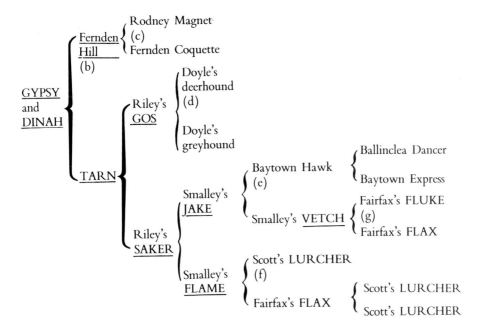

Notes: (a) Names in capitals indicate lurchers; names underlined indicate dogs illustrated in this Chapter

(b) Fernden Hill, semi-finalist Waterloo Cup as puppy, broke hock in 4th round Waterloo Cup in following year when leading the eventual winner, Minnesota Miller

(c) Rodney Magnet won Waterloo Cup 1970 (this breeding is not extended as it is "in the book")

(d) These two dogs were pure-bred but details are still not available

(e) Baytown Hawk, a track dog, ran mainly at Newcastle where he held several track records (again, the pedigree is not extended as it is "in the book")

(f) "Scott's lurcher"; though several people in the Kings Lynn area remember these dogs it has not been possible to find out their names

(g) FLUKE was "by a shepherd's lurcher out of a bitch at Brough who was bred by the Buxtons in Herts".

gone and the puppy is one of a pile with nothing to distinguish them except colour. I have no idea why this can happen; I know nothing about horse-breeding but I am told that it can sometimes happen with bloodstock. But if one has the eye to see it, the wet puppy that looks like a goer should be noted. It is not always the one that is kept; I well remember a whippet litter in which one bitch puppy had, for the first few moments of life, everything that one hopes for: a certain coursing winner if I have ever seen one. When the bitch had finished whelping I left her for an hour and on going back I found that she had lain on two puppies; one of them was, inevitably, the puppy I had noted.

Lurchers, like greyhounds, can produce very big litters. I think that six is quite enough for a bitch to rear properly and I would put down any above this number. The choice has to be made quickly; I would first of all put away any puppy that appeared weak in any way or did not try and suck the moment it was born. Life can be tough for lurchers and if they do not struggle at the beginning they probably will not struggle later on. The only suggestion I make is on colour and sex. I have a prejudice – quite groundless, probably – against white dogs; and as most people prefer bitches for working, a bitch puppy should not be put down without a good reason. Let the dogs go first.

With the mixtures of blood in so many lurchers, some experienced breeders do not expect more than 25% of a litter to be any good. Certainly there are a lot of lurchers to be seen about that no conscientious breeder should ever have reared to maturity, but unless there is something definitely wrong it is difficult to cull a litter between birth and three months; and by three months, those that are going should have been sold. (If they haven't been sold by the time they are 12 weeks old the breeder will start to lose money on them and the amount can accumulate very fast.) Culling is a problem, but I would not go quite so far as to recommend the methods of a West Midlands character, well known in the coursing field, who reared his litters of greyhound puppies in a large bullock yard where they had shelter, straw bedding, water, and room to move around. Two or three times a week a carcase or some offal was thrown in to them, over which they would fight and, as greyhounds can fight, sometimes to the death. Dead or badly wounded puppies were removed and at the end of six months 'rearing' the breeder was left with, possibly, only one dog, but that one was definitely the survivor of a very tough upbringing.

If the dew claws are going to be taken off this should be done about

Associated Newspapers Group

"FERNDEN HILL" ("Dick") about to pick up his hare in the Waterloo Cup. His running weight was 82 lbs

PARENTS

"TARN", by "GOS" out of "SAKER": 25 inches at the shoulder and 58 lbs in weight. One of her litter-brothers, "PADGE", is shown on page 15

Author

"GOS"; a deerhound-greyhound cross. A very fast dog and a natural jumper (see page 156

GRAND-PARENTS

"SAKER", a Norfolk lurcher, by "JAKE" out of "FLAME". A bitch with tremendous stamina and a determined killer

the third or fourth day. I have no strong feelings about removing them from lurchers. Coursing greyhounds have them left on, coursing whippets have them removed. Some hounds have them left on, some have them removed. I have often heard it said that without dew claws, dogs have difficulty in turning sharply on wet ground; this may or may not be so. I've never seen a whippet having any difficulty and whippets will turn very much more sharply than any other coursing dog. On the other hand I have seen lurchers with blood pouring from their dew claws after running on short, stiff stubble; if this happens and the dew claws have to be taken off an adult dog it means a proper operation, with the dog laid up for some weeks. On balance, I prefer to have them off but a lot of people will disagree with me.

Rearing
The facilities needed for breeding lurchers are no different from those needed for any other sort of large dog; a secluded place where the bitch can whelp in peace and quiet and where the puppies can be left for the first four or five weeks. They merely need room enough to squirm and stagger around, and the bitch should keep them and the whelping box clean. Dosing for round worms should be done by the fifth week at the latest with the second dose a fortnight later. Some knowledgeable breeders say that the earlier this is done the less check there will be to the puppies' growth. I like to get puppies out of doors at five weeks if possible, even if it is only during the day-time, for an hour or two. But by six weeks I like to have them outside for good unless I have so mismanaged affairs that the litter has been born in mid-winter.

Over the years, I have found that the simplest form of housing that combines size, warmth, freedom from damp and draughts, and cheapness, is a straw bale house. About 21 bales will make three sides of a house, three bales high, with a slight projection on the fourth side. This side is made of timber with half opening as a door. A tongue and groove timber ceiling is fitted, over which goes some form of insulation: polythene sheet and straw, or plastic sacks filled with straw. On top, is a corrugated-iron sheet roof with a projection, four or five feet beyond the door, which forms a porch under which the puppies can be fed in wet weather. By moving the wire-netting fence further and further out until they have the freedom of the whole run or paddock, such a house will do for a litter of lurchers until they start to go at ten to 12 weeks, and will then house the ones that are left, if necessary until they are a year old. The straw bales are then either burnt or composted and a new house is built for the next litter on a different site. Inside the house I put a wooden floor, some six inches off the ground, and the puppies have a couple of bales of straw as a bed. With this warm bed and warm house they can be left outdoors in the worst of weather and come to no harm.

Feeding

Feeding the lurcher is no different from feeding any other working dog; or at least it was no different when one could get meat fairly cheaply. (What ages ago that seems to be!) Now one has to get what one can and no source of meat can be despised.

A good lurcher should be capable of feeding itself, given the opportunity to do so. Many lurchers try to feed themselves whenever an opportunity offers which can lead to trouble. Until the puppy learns some discipline, nothing edible is safe unless it is in a cupboard or placed at least seven foot up on a shelf. Perhaps remedies are not quite so drastic now as they used to be. Parson Woodforde records in his Diary, on April 11th, 1794: "One of my greyhounds, young Fly, got to Betty Cary's this morning and ran away with a shoulder of Mutton undressed and eat it all up. They made great lamentation and work about it. I had the greyhound hanged in the evening." However, in these crowded times and with meat at the price that it is, stealing legs of mutton should, perhaps, not be encouraged.

The dog's teeth and digestive system show that he is a meat eater. His teeth are not designed for masticating but for tearing off large lumps of flesh and his digestive system can cope with large amounts, taken in a short time at long intervals. The wolf and the wild dog run down their prey, strip the carcase, leaving only the largest bones, and trot home to their dens to sleep off the meal; they will not hunt again until they are hungry. Though the modern dog has little or no opportunity of copying his wild relatives, his teeth and digestive system are the same as theirs and his requirements are the same: a fresh carcase, guts, meat, skin and bone, to tear at once every day or two. And, to follow the hunting analogy, a fast day once a week. The meat will provide the proteins and the half-digested green stuffs in the stomach and intestines of his prey provide his other requirements.

But very few dogs today ever get their teeth into a carcase of any sort; and, outside hunt kennels and coursing dogs, not many modern dogs would know what to do with a carcase if they were given one. A great

"JAKE"; in old age. By a record-breaking track greyhound out of a Norfolk lurcher

GREAT- and GREAT-GREAT-GRANDPARENTS

"FLAME" and "VETCH" who go back to the Kings Lynn dogs

Jim Meads

"TIGER" jumping in public

"MOST LURCHERS WILL
JUMP FOR THE FUN
OF IT..."

"GOS" jumping at home

Author

The end of the day

Jim Meads

NOT MANY HARES IN KENSINGTON: Mrs Riley and the author with "TI" and "TARN", litter brother and sister, taking part in Crufts Personality Parade, Olympia, 1978. Another of this litter, "PADGE" is shown on page 15

many dogs are fed from tins or packets; for the majority of dog-owners it is very much easier to open a tin or shake out pellets from a bag. But such feeding is very expensive and I personally do not see the point of contributing to a petfood manufacturer's Rolls-Royce and holiday in the Bahamas (I understand that the U.K. market is worth more than £200 million a year retail). I live not too far from a butcher who slaughters his own meat and I collect two or three beasts' paunches once a week; in winter these will hang in the feed room whilst being used but in the summer I cut them up into daily amounts and put them into the freezer. All that has to be done then is to remember to take out the next day's food in time. For mixing with chopped paunch, or for a change of food, or for emergencies, I keep a supply of dried meat (used by soaking with boiling water and feeding when cold), and biscuits made by Roberts & Co., of Dunchurch, Rugby. I also keep a supply of meal made by Wilson's of Dundee; the dogs like it and I find it invaluable for tired dogs after a long day's work or for putting flesh on a dog that has lost condition. It is, of course, a complete food by itself.

Choosing a Puppy
Not every lurcher-owner wishes to breed his own dogs; many do not have the facilities. For these people, buying-in has to be done. The first question is: what age should the dog be that one looks for?

The temptation is to save time and buy an adult dog; the danger here is that adult dogs that are any good are not for sale except in rare cases, when perhaps an owner has suddenly died. If I am offered an adult dog I want to know why it is for sale – I have not, in fact, bought an adult for some 30 years. Good lurchers are like good sheepdogs or good gundogs; a lot of time and trouble has been taken over their upbringing and they are not lightly got rid of.

The person in search of a lurcher should mark down a bitch that he has seen coursing or working; should find out when she is going to be bred from; should try and have a look at the dog that she is going to; and, if he likes them both should waste no time in booking a puppy. This is the first step. The next, and much more difficult step is to choose a puppy when the litter is ready to go, at about eight to ten weeks. It is difficult enough for the breeder who wants to keep one for himself; but he will have seen the litter daily and will, if he knows the old coursing trick, have hung them up by the tail at a week old and measured them on the stable door. Other things being equal, he will

159

keep the longest puppy for himself; it will, most likely be the fastest. Lurchers grow so unevenly that make and shape is not much of a guide until they are almost grown up. I would avoid a three-month-old puppy that looked fine or pretty; it will amost certainly grow up a weed, whereas the coarse puppy will usually fine down. Temperament is important and I would avoid the puppy that stays by the kennel door or goes back inside on sighting a stranger. The first puppy to move or the one that comes up to the wire to greet a stranger is the one to look at. Most lurchers are friendly dogs and a nervous one should be avoided; it is unlikely to get much better as it grows up and it will always be a nuisance to train.

CHAPTER FOURTEEN

Training

There are many stories told of the intelligence of lurchers and the wonderful things they have been trained to do; given a clever dog – and many lurchers are very clever indeed – and given time and patience, there is probably no dog work of any sort that a lurcher cannot be taught to do. But when one thinks of everything that the 'complete lurcher' should be capable of doing: hunt, jump, catch kill and carry, drive to a gate net or long net, stand and mark to ferrets, point, mark and retrieve game, both fur and feather, steal and carry eggs, warn of the presence of strangers, and guard the house or van amongst other things, one realises that 18 months is not too long to train a dog.

I do not pretend to be a dog trainer beyond elementary matters; but I did once mystify an acquaintance by showing him some dogs that would 'meet me after dark'. I sometimes take dogs out round a few meadows, last thing at night, when there is enough moon to see by; on this occasion the moon was on the wane and I had been out for several nights running, so by 10.30 p.m. the dogs were getting impatient. I suggested a walk to my guest and, whilst he was putting on his boots, I let the dogs out of the house. By the time we started out they had disappeared and I guessed that they had gone to a patch of thorn where they had caught a rabbit that morning. We walked over a couple of fields, not seeing any dogs, and avoiding the patch of thorns. On arriving at the far gate I blew a 'silent' whistle, turned up to its highest pitch, hoping that my guest, who was a bit deaf, would not hear it. Within a couple of minutes the five dogs had joined us at the gate, much to his surprise, and we walked a little way down the lane and back to the house discussing the higher training of dogs; I was hard put to it to answer some of his questions. It was, of course, an easy trick to play, and the higher training of lurchers is a job for the few experts still at the game. Those who use gate or long nets and who take game by other quiet and methodical ways are fully capable of training their dogs to these sometimes frustrating and often hair-raising sports. But even for the lurcher-owner who does nothing in particular with his dog there are one or two essentials; for the lurcher is the product of a great many generations of dogs that have been bred to run after and kill anything that runs away from them. Such a dog is lethal and it can be a public nuisance unless it has had some discipline instilled.

The dog that can be recalled when after a hare or fox is either very highly trained or it has not got its heart in the business; but otherwise, a dog must come back at once on command. The dog that will walk to heel and show reasonable obedience is a much more pleasant companion than the unruly dog. For those who live in the countryside the rule is that their dogs must be absolutely steady to all farm stock at all times. To walk past sheep or cattle without looking at them is not enough; when a hare gets up, one never knows where the course is going to end and one must be quite sure that one's dogs will return across country, passing farm stock, without any possibility of trouble. "By kind permission of the landlords and tenants" does not only apply to public coursing though it is only on Club cards that the phrase appears. In these socialist days, of anti-rural legislation and financial stringency, some bullocks chased through a fence, or the merest suspicion of sheep-running can strain the generous goodwill that is still so freely given by so many farmers. The damage done by one unruly lurcher will rub off on the rest and good running grounds can be lost through the stupidity and idleness of one man and his dog.

I have mentioned an 'Acme' silent whistle; this, to my mind, is essential to the lurcher-owner. Recalling dogs by shouting, no matter how dulcet the voice of the caller, not only disturbs game far and wide but also advertises one's presence when it is, perhaps, more tactful not to do so. A dog that is used to an 'Acme' whistle can be recalled from seemingly incredible distances without disturbing the countryside; and the different pitches of the whistle can be used for more advanced training.

A lurcher must be able to get across country without assistance. A dog that dithers when his hare goes through a hedge loses too much ground to get on terms again; and there is nothing more infuriating than the dog that cannot keep up across country by jumping, creeping or going round and catching up. Most lurchers will jump for the fun of it but jumping is also a question of confidence which can be built up in early life. By the time puppies start playing and chasing each other in earnest, say at about four months, their run should be divided by a low barrier – not more than a foot high – made of sleepers, scaffold planks, etc., with one or two straw bales as well. The puppies will take these obstacles in their stride without thinking about them and the barrier can be gradually raised as the puppies grow. Eighteen to 24 inches is quite high enough for young bones and joints but, provided that they are not overfaced or damaged when they first go out into the countryside, it should never occur to lurchers that they cannot jump. Similarly on early walks, the opportunity should be taken of crossing hedges and fences so that the puppy has to follow and work out for himself how to creep and squeeze. There is only one sort of fence that gives me palpitations if I see my dogs going for it after a hare and that is the hedge with a bullock wire a yard out in the field beyond it. The dog that flies this type of fence often gets caught on the wire as he lands and one can only hope that he sees the wire in time and banks the hedge.

An oft-debated subject is what age the lurcher should start work. Some say as early as possible, get him after a hare at six months and if he's no good get rid of him. Others say wait till he's a year old. In my opinion it depends entirely on the dog's breeding and conformation. If he comes of heavy stock, is well boned, is carrying a lot of flesh and shows signs of more growth to come, he must be given as much time as possible before he is let loose after a hare; too much work too early will build up muscle and this muscle will pull the bones, that are still forming, out of shape, the muscles and tendons being stronger than the bones they are attached to. If the puppy comes of lean and lightish stock, is an early developer, if he doesn't sprawl when playing with the other dogs, if he jumps or climbs out of the run at an early age then, perhaps, not so much harm will be done by some early running.

The important thing is that the puppy should never be given more to do than his development can cope with and it should never occur to him, until he is grown up, that he cannot catch what he is after. Assuming that he is a spring or summer puppy, he should do no more for his first winter than play with other dogs, and taken for short walks in neighbouring fields to get him used to finding his way through hedges and fences, and knowing the way home. Longer walks should be mostly on a lead, particularly where hares are likely to be found. Equally so with rabbits, as there is nothing so off-putting to a young dog than finding that whatever he chases will disappear down a hole; early experiences of missed rabbits lead to the dog that hesitates when he should dive to pick up a hare. Give him all the road walking you can spare time for from spring onwards, to harden his feet, joints and muscles; and if you can find some 'myxy' rabbits, let him catch a few to see how easy it is. Then, when the corn is cut, take him out very early in the morning, at first light, with an experienced dog to make up the pair; pick some small fields with long stubble and slip the dogs right on top of a hare. No matter how short the course, don't be tempted to repeat it but take the puppy home whilst he is pleased with himself and dancing about for more. Gradually extend the lessons and by the time the winter comes he will be fit and hard and full of confidence and ready to take his place in your team. It is all too easy to damage a puppy beyond repair by one unthinking action.

Training for Fitness

These thoughts are partly the result of my own experience with running dogs and partly what I have gleaned from greyhound trainers who have been kind enough to pass on some of their hard-won knowledge; and, since a good lurcher must have more or less of the greyhound in its make-up, training methods for lurchers can best be based on those methods that have been found effective with greyhounds.

Adair Dighton's words on the principles guiding the greyhound trainer are as true now as when they were written in 1921:

"In the training of anything, man, horse or dog, the whole secret depends on an intimate knowledge of the subject under training. What is one dog's meat may be another's poison, and to be successful as a trainer it is absolutely necessary to know and understand each dog's eccentricities and whims. Some dogs do well on horse flesh: others do not. Some do better when kennelled with a companion: others prefer to be alone. Some are delicate feeders and must be tempted: others are worse than the proverbial pig and must be restrained. All these and a thousand and one little things must be noticed, remembered and understood; otherwise the game is better left alone. The man who feeds and trains half-a-dozen greyhounds all on the same food and each the same amount of exercise may occasionally, by good luck, have one fit, but he is not a real trainer."

Of course the lurcher has not got to be trained to the minute as the greyhound has to be for a particular stake; but unless a lurcher is hard and fit before he starts work he will inevitably damage himself, perhaps for good, so the effort of getting him fit is not only worthwhile but is essential. Starting from scratch, with a dog that has had a quiet summer — and has not been run to skin and bone with lamping — the work will take a month or six weeks, depending on how soft the dog is and how much time you can spare each day.

To start with the dog must be clean, inside and out; "all dogs have fleas and worms" is a sweeping statement but is more often true than not. So go to your vet, tell him the weight of your dog and get whatever tapeworm medicine he recommends and also a body-wash to get rid of external parasites.

Now comes the exercise and, contrary to what many beginners think, it is not galloping that gets a dog fit but it is steady walking on hard roads. If the dog is fat and has soft feet, three miles a day for the first week is probably enough. For the second week the target should be five or six miles a day. The distance should be done in about two hours and when you get back home you should rub the dog down. This "rubbing down" is the equivalent of strapping a horse and should be done just as systematically: start with the neck muscles and go on to the shoulders, back, ribs and fore-legs; then turn the dog round and massage the loins, quarters and hind legs. It can be done with bare hands — and obviously, if you are using an embrocation you will put it on with bare hands — but a hound-glove gives better results, especially with a rough-coated dog. Inspect the dog's feet and if there is any soreness (usually caused by too much road work, too fast, too soon), bathe the feet in salt water, or apply a mixture of Friars Balsam and Myrrh, and ease off the work for a day or two.

At the beginning of the third week the proper work will start and though the length of your daily walk need not be increased the speed must be: try and do the distance in an hour and a half. If you can spare the time, increase the amount of daily hand-rubbing and don't forget to look at the dog's feet on getting in from exercise. The hard road should keep his toes short but watch for any nails that are not being worn down — a sign of a bad foot — and keep them clipped. As the work increases so the dog's food must contain more protein and less carbohydrates; if the weather is hot and the dog is fat, add some salt to his food to replace food to replace the salt lost in sweat.

A little fast work can be started in the fourth week but this should not be hurried. Remember that, unlike a horse whose speed can be regulated by the rider, a dog will go flat out; and the unfit dog that gallops flat out is going to do itself an injury of some sort. The daily

165

road work and hand rubbing are what gets a dog fit and gallops are used to sharpen it up. The best way to give a dog — or a number of dogs — a gallop is for a helper to hold them whilst you, the owner, go to the far end of a meadow, preferably one that slopes uphill. For more than two dogs your assistant should wear a stirrup leather round his waist to which all the leads can be fastened; this leaves his hands free to undo each collar in turn. Dogs should be loosed singly and at sufficient intervals to prevent them catching each other up and, probably, fighting.

For the third fortnight the road work can be increased in speed and, if possible, in distance; and the dogs can have a daily gallop or two so long as they are on their toes. But watch for any sign of staleness; if it does appear, cut out the fast work and give the dog concerned a day's lounging in place of road work. As a change from galloping, try a strong fishing rod and line with a rabbit or hare skin on the end. Swinging the skin round and round and across the circle will make a dog start and stop and turn and will exercise all his muscles. Whippets and small lurchers can become particularly sharp about catching the skin but watch out with big dogs unless you are large yourself; a 60 lb dog between the shoulder blades will knock most men over.

Remember that the object of training is not to produce an unnatural condition but to produce a dog that is somewhere near the peak of physical fitness and it is the balance between food and exercise that produces this peak. You have two aids to help you, the weighing machine, and your hand and eye. Dogs will vary but a rough guide is the fact that many dogs' best running weight is the weight they were as a healthy, year-old puppy. As the work proceeds there should be a firmness of muscles, a spring in the step, a shine on the coat and a general air of well-being (don't forget that this will apply to you too if you are doing six miles a day with the dogs!). For a final check, put your hand on the dog's chest muscles: if these are firm, you are on the right lines in your training.

CHAPTER FIFTEEN

First Aid

The days are long past when the only veterinary help came from the local cow doctor and when the dog-owner had to be guided by his bookshelf in the recognition of a host of diseases, all of which were written up in great detail; he then had to decide which prehistoric remedy to apply. For instance: "A mild remedy for worms, unattended by any danger: take of powdered glass, as much as will lie on a shilling, heaped up; mixed with butter and given as a bolus and to be followed with castor oil after six hours." The powdered glass no doubt settled the worms but what did it do to the dog's inside?

Diseases of the dog are now the province of the vet. The diseases that, not so long ago, were accepted as part of the life of a dog and which decimated hound kennels, are dealt with by routine injections and inoculations and the dog-owner that does not have his dog inoculated against hard pad and jaundice at about 12 weeks of age has no business keeping dogs. The healthy dog has a bright eye, a cool damp nose (though this is not an infallible test), a good appetite, is lively and has an overall look of well-being. Any sign of listlessness – as opposed to tiredness — with loss of appetite, raised temperature, (a dog's normal temperature is about 101.3° F), vomiting or diarrhorea and the dog should be put on its own in a warm place and kept on honey and water for 24 hours. If the signs continue the dog should be taken straight to the vet. Lurchers are normally very hardy dogs and, given their quota of exercise, food, and rest on a dry bed out of a draught, will remain fit and well into old age. There are, however, certain occupational hazards which occur in the course of a longdog work. These are cuts, tears and staking, strained toes and legs and, in the summer, sore eyes from hunting through nettles.

Cuts and Tears
These are mainly caused by barbed wire. If the cut is on the leg below

the elbow or hock, or on the ribs or back and is skin-deep, it probably need not be stitched. Green oil and the dog's tongue will deal with it. If simple stitching is needed and it can be done soon after the accident most dogs will stand it without a local anaesthetic. But if the tear or cut is more than skin-deep, or if it is on the neck, chest, stomach, or inside the elbow or stifle it must be stitched and unless the owner is an expert it must be stitched by a vet. Get the dog to the surgery as quickly as possible. A punctured wound must be kept open as it heals and, again, Green oil is useful if old fashioned. Cut and damaged feet should be dealt with by a three-parts to one-part mixture of Friars Balsam and Myrrh. If nothing else is immediately available, cold tea makes a good eye wash but Optrex will sooth a dog's eyes if he has been into nettles after a rabbit. The other use for cold tea is washing out a dog's mouth after a course; a small plastic bottle of very milky tea or plain tea and honey is easily carried and clears the froth from a dog that has been hard run.

Strains

A longdog has to work under all sorts of conditions, grass, plough, stubble, root crops, stones, flints, mud and woodland. He very seldom has the luxury of the running ground on which the greyhound is slipped, and he may well have to run on ground that is hard from frost or drought. Provided that he is given enough road work in the autumn to keep his feet and joints hard, damage should not occur too frequently but if a foot or leg is strained or a toe is knocked-up the only cure is rest. Strains can be reduced by various means; Radiol is an old stand-by, Green oil, though old-fashioned, will work just as well as many more modern embrocations, cold water is not to be despised; but the one absolute essential is rest. A dog can be given a pain-killer and can run again after damaging itself; but such action can only make the damage worse and with lurchers, where there is no Waterloo or Barbican Cup at stake, it is an extremely stupid and selfish thing to do. For any strain damage, rest the dog and do not let him out of a walk until he is going absolutely sound again. If a chance is taken the end result will almost certainly be that the dog will break down for good.

Drinking Water

A warning about letting dogs drink from stagnant water. Very few lurchers are as highly trained as coursing greyhounds but they are often expected to run two or three courses in a morning; and these courses can be very long indeed. Some lurchers are over-run and are skin and bone, others have some weight to lose and it is these dogs that want to drink during the day's work.

Most lurchers' work is done on farmland; nowadays an enormous amount of chemicals are used on nearly all farms, some of which are for stopping things growing, some for starting things growing, some for killing bugs and beasties, some for dealing with mildew, etc. They are safe when diluted but none of them are designed to do a dog's inside any good. A usual method of application is for plastic sacks to be taken out on a trailer and the distributor filled from the sacks. Where the trailer has stood it is possible for a concentration of spilled chemicals to build up which may last for a long time. If a dog drinks water from a wheel rut where such spillage has occurred the trouble may not show for some days by which time the incident may have been forgotten, or even not noticed, and diagnosis and treatment may be difficult.

It should be a rule, when away from home, never to let a lurcher drink except from running water or from a cattle trough; or, better still, only let him drink when he gets back to the van where you have, of course, got a bottle of water and a bowl for him.

First-aid Bag

For the lurcher-owner who does more than an absolute minimum of work with his dogs, a first-aid bag, kept filled and put into the car or van whenever the dogs go out can be the saving of a dog's soundness and possibly of his life. An army haversack will hold all that is necessary, and I suggest, for the contents: surgical scissors, tweezers, a curved needle and gut or thread, cotton wool, a one-inch and a two-inch bandage, a bottle of saline solution (one teaspoon salt to a pint of water), Green oil, Friars Balsam, a packet of gauze and some ointment (Acriflex or Savlon are useful). Like the clasp-knife that I always carry, the first-aid bag may not be needed for week after week; but when it is needed, it is needed urgently, and it is not much help knowing it is 20 miles away at home.

A Dignified End

Lurchers will normally lead long and healthy lives and, unless a foot or joint is irreparably damaged, will run up for most of their days; as speed gets less so cunning becomes greater. But, as with all living things, the time comes when the body begins to fail, and every owner must remember that when taking on a lurcher – or any other dog for that matter – he or she takes on the responsibility of seeing that its latter days are comfortable. The dog will need less food as it gets older and, given its freedom, will take as much exercise as it wants. A few dogs die a natural death; many have to be put down. Where there is pain, disease, or the inability to perform natural functions, the decision is straight-forward. It is the slowly deteriorating dog that presents problems. It can easily happen that the person who is with the dog every day is the last one to see the real change in condition, but it is only the owner who knows the dog in all its moods who can tell when the time has really come.

No dog should be kept alive on drugs just to suit its owner's feelings. Modern methods of euthanasia are painless and very quick but in the clinical atmosphere of the vet's surgery the old dog needs your reas-surance. Take him to the vet yourself and stay with him until the injection is done. After all he has done for you, the very least you can do for him is to see that his last moments are dignified and unworried.

"On your grave beneath the chestnut bough
Today no fragrance falls nor summer air,
Only a master's love who laid you there
Perchance may warm the air 'neath which you drowse
In dreams from which no meal-time calls may rouse,
Unwakeable, though close the rat may dare,
Deaf, though the rabbit thump in playful scare,
Silent, though twenty foxes screech their vows.

"And yet mayhap, some night when shadows pass,
And from the fir the brown owl hoots on high,
That should one whistle 'neath a favouring star
Your shade shall canter o'er the grass,
Questing for him you loved in days gone by
Ere death, the dog thief, carried you afar."

Bibliography

I did not, originally, intend to add a bibliography to "Lurchers and Longdogs"; lists of books which the author has consulted belong to more erudite works than this. But I have often been asked for the names of books on lurchers and allied subjects so I hope that the following list may be of use to those who want to read more. Unfortunately many of the books I mention are out of print; for those with access to one of the big libraries — British Museum, Fitzwilliam, Bodleian, London etc — the answer is easy. Others have to keep an eye on the sporting shelves in second hand bookshops; shelves which have shrunk alarmingly in the past ten years or so. I wonder where all the sporting books have gone?

The early use of running dogs
I imagine that if one could read Egyptian hieroglyphics one might learn something about coursing with saluki type dogs under the Pharoahs. But for practical purposes the earliest book that is worth reading is "The Master of Game", written by Edward Duke of York in about 1410. Most of the book is a literal translation from Gaston de Foix' "Livre de Chasse" written some twenty years earlier. There is no need to worry about the language of "The Master of Game" as an excellent edition was produced in 1911, edited by William Baillie-Grohman and published by Chatto and Windus. From here on — 1410 on that is to say — there is a series of books on sport and hunting, most of them copied from each other; they are only worth reading from a scholarly point of view. The next book of any practical value is by Dr Caius; his "Englishe Dogges" was written in 1570 and I have quoted briefly from it on page 8. Not until 1760 do we find anything of the lurcher when Thomas Fairfax wrote "The Complete Sportsman". He also has some interesting remarks on the "tumbler" which, from his description, I would imagine was a whippet. In 1804 William Taplin wrote "The Sportsmans Cabinet" which has detailed descriptions of sporting dogs including the lurcher and the "cur" or drovers dog.

Lurchers

There are no early books on the lurcher itself; indeed, "Lurchers and Longdogs" was the first book to be written solely about lurchers. But Phil Drabble has a chapter about lurchers in his "Of Pedigree Unknown" (chapter seven) published by Cassels, 1964 and Brian Vesey-Fitzgerald has something to say about them in "Its my Delight" (Eyre & Spottiswoode, 1947). Cyril Heber-Percy's "While others Sleep" is a well-observed story of a poacher and his dog but I would query the dog's height; 36 inches is big for any longdog!

Greyhounds

There are a lot of books about greyhounds; some good, some mediocre, some merely repetitive; most of them include something about coursing. In chronological order I would recommend "Stonehenge" on "The Greyhound" (Longman, Brown and Green, 1853). Though out of date for most practical purposes now, Dr Walsh was a member of the first National Coursing Club Committee set up in 1858 and many of his comments are still worth reading. The next book which I would strongly urge any courser to try and buy is Adair Dighton's "The Greyhound and Coursing" (Grant Richards, 1921). His methods of training are certainly not out of date. Finally, from that walking encyclopaedia of greyhounds and coursing, the late H. Edwards Clarke's "The Greyhound" (Popular Dogs, 1965). Besides being a most charming man, Eddie Clarke had forgotten more about coursing than the rest of us will ever learn. For anyone interested in the bigger kennels and coursing meetings of the 1860's and earlier the various books by "The Druid" are worth looking for.

Sheepdogs

There are many books about sheepdogs, some factual some fictional. For two factual books, try A. L. J. Gossett's "Shepherds of Britain" (Constable, 1911) and John Holmes' "The Farmer's Dog" (Popular Dogs, 1970). There are some wonderful descriptions of sheepdogs working in Ernest Lewis' "Beth" (Constable, 1934) and his "The Hill Fox" (Constable, 1937).

Hounds

There can be only two choices here: Sir John Buchanan-Jardine's "Hounds of the World" (Methuen, 1937) and Daphne Moore's "Book of the Foxhound" (Allen, 1964). I would, however, add Jack Ivester-Lloyd's "Hounds of Britain" (Black, 1973); the Commander knows his subject.

Terriers

Terriers do — or certainly should — interest the lurcher man and books about them are legion. But there is still only one complete work on terriers and that is Sir Jocelyn Lucas' "Hunt and Working Terriers" (Chapman and Hall, 1921).

Deerhounds

I prefer George Cupples' "Scotch Deerhounds and their Masters"; now, unfortunately rare and expensive. A. N. Hartley's "The Deerhound" (privately printed at Peterborough, 1955) is short but good.

Saluki

The best book I know of is David and Hope Waters' "The Saluki in History, Art and Sport" (David and Charles, 1969); Mrs Waters not only has some very high class show dogs but courses her salukis regularly and with great success.

Borzoi

Winifred Chadwick's "The Borzoi" (Nicholson and Watson, 1952) but make certain your copy has the extra section on the Perchino Hunt by David Walzoff.

Whippets

There are various books on whippets; as a whippet owner and courser I disagree with some of what he says but the late C.H. Douglas Todd's "The Popular Whippet" (Popular Dogs, 1961) is still the best to my mind. I have written my comments on his book in the whippet section of the definitive book "Coursing".

Coursing

The first, complete, definitive book on coursing is "Coursing" (Standfast Press, 1976). It is written by a number of people, each knowledgeable in his sphere, and covers every possible facet of the sport. It will become a collectors book. Anyone interested in coursing should read the British Field Sports Society's investigation, "A Review of Coursing" by Owen

Stable QC and R. M. Stuttart (B.F.S.S., 1971). It is usually referred to as the Stable Report. There is much about coursing in the greyhound books that I have recommended. A book that is somewhat dated from the technical point of view but is still worth reading is Harding Cox' "Coursing" (in the Badminton Library series, 1892). I have already quoted some extracts from William Scrope's "Art of Deerstalking" (Edward Arnold, 1897, though there are various different editions). Charles St.John's "Wild Sports of the Highlands" (Gurney and Jackson, 1927 but again many editions) is also good on deer coursing. Finally there is the story of coursing deer at Atholl which I have quoted from Sir Samuel Baker's "Wild Animals and their Ways" and a very good description of coursing deer in Cumberland can be found in Ernest Lewis' "Beth", already mentioned under Sheepdogs. Those interested in the fate of coursing should read the Minutes of Evidence and Proceedings of the Select Committee of the House of Lords, Hare Coursing Bill (H.M.S.O., 1976) if for no other reason than to see how staggeringly ignorant the various "anti" groups are of basic country facts.

Gypsies

For an academic study of Gypsies one must read the Gypsy Lore Society Journals. These run from the 1880's to the present day but, to me, the interesting thing is that in all these 90 years of scholarship there is hardly a mention of dogs or horses. The clever men who studied Gypsies obviously had their minds on higher things. Essential reading is Brian Vesey-Fitzgerald's "Gypsies of Britain" (Chapman and Hall, 1944 but I also have a later, paper-back, edition), and there is much of interest in C. H. Ward-Jackson's "The English Gypsy Caravan" (David and Charles, 1972). For life on the roads I would recommend one of Dominic Reeve's books; perhaps "Smoke in the Lanes" is the best.

Rabbits

For wild-life in general one should have a copy of Courbet and Southern's "The Handbook of British Mammals" (Blackwell, 1977). For rabbits I can only recommend R. M. Lockley's "Private Life of the Rabbit" (Andre Deutsch, 1964).

Hares

Without a doubt the best insight into the mind of a hare is in Chapter Two of W. Lovell Hewitt's "Beagling" (Faber and Faber, 1960). It is well worth reading by anyone who, like myself, finds he knows less and less about hares the more he chases them. There is a lot of good stuff about hares as well as some very good stories about private and public coursing in the Fur, Feather and Fin series "Hares" (Longman, 1896).

Foxes
Brian Vesey-Fitzgerald's "Town Fox, Country Fox" (Deutsch, 1965) is worth having on one's bookshelf; and Richard Burrows' "Wild Fox" is a wonderfully detailed investigation into the habits of a small group of foxes in Gloucestershire (David and Charles, 1968).

Deer
For those who want to know, G. Kenneth Whitehead's "Deer of the World" (Constable, 1972) is the best. For local identification, the "Field Guide to British Deer" (Mammal Society, 1957) is perfectly adequate. Otherwise, see the various B.D.S. pamphlets.

Longdogs overseas
I have already mentioned two of Sir Samuel Baker's books. Apart from these it is really a question of what one can find as there are very few lists of books published overseas that help. From America I very much enjoy reading Leon V. Almirall's "Canines and Coyotes" (Caxton Printers, Idaho, 1941) which has racy descriptions of coursing coyotes in the 1920's and '30's. Two very recent books are Steve Copold's "Hounds, Hares and other Creatures" (Hoflin, Colorado, 1977) which I found unnecessarily sadistic; much more enjoyable is M. H. "Dutch" Salmon's "Gazehounds and Coursing". One must remember that they are writing for a different public who have not our traditional approach to coursing.

The Game Laws
For the early game laws one has to read John Manwood's "Treatise of the Forest Laws" (various editions from about 1570) but the source book is G. J. Turner's "Select Pleas of the Forest" (Seldon Society, 1901). For the more up-to-date law the best is "Oke's Game Laws" (mine is the fifth edition, Butterworth, 1912) and "Notes on the Game Laws" by "Woodman" (Herbert Jenkins, 1962) is good.

And finally First Aid
Damage control is really the province of your 'Vet but it is very

worthwhile learning something about how your dog is put together and how it functions. Sisson/Grossmith "Anatomy of the Domestic Animals" (Saunders, 1969) will teach you the anatomy of your dog and the diagrams can be a great help when trying to pinpoint damage. Jones' "Animal Nursing" (Pergamon Press, 1972) is the text book for R.A.N.A. and the "T.V. Vets Dog Book" (Farming Press, 1974) has some down-to-earth photos but is a bit short on the ills that beset running dogs. Finally, H. Montagu-Harrison's "The Greyhound Trainer" (Cashel Press, 1962) taught me a lot about running dogs that I should very well have known.

INDEX